Climate Perspectives from the Congo Basin

This book considers the global question of climate change from local perspectives in the context of Central Africa.

Bila-Isia Inogwabini examines attempts made by the international community to respond to the global challenges posed by climate change in the Congo Basin and highlights that these attempts have so far produced limited results. Abject poverty and the lack of academic, technical, institutional and governance capacities have made it difficult for these solutions to take root in local conditions. Taking a novel perspective, Inogwabini argues that what is needed is not austerity in the use of natural resources but rather increased material affluence for these communities, which will enable individuals to create their own ways to survive through the tides of climate change. He considers factors including social inertia, climate skepticism and lack of political structure and presents a climate change action plan that is targeted at the local level in the Congo Basin.

Overall, this volume will be of great interest to students and scholars of climate change, global development and African studies more broadly.

Bila-Isia Inogwabini teaches Biodiversity, Climatology and Climate Change and Ecology at several academic institutions in Democratic Republic of Congo (DRC). He currently heads the Department of Environment and Renewable Resources Management at the School of Management of the Catholic University of Congo (Kinshasa) and has been a visiting scholar (external resource) at the Department of Aquatic Sciences and Assessment at the Swedish University of Agricultural Sciences (Uppsala, Sweden) for a decade. He is a top-performing Conservation and Sustainable Development Manager with over 29-year-long field experience over which he led many conservation, humanitarian and sustainable development projects and programs across Central Africa (Cameroon, the Republic of Congo, Central African Republic and the DRC). His key areas of expertise include biological survey designs, biodiversity and ecology, freshwater and fish ecology, climate change, sustainable development, quantitative and qualitative research methods and wildlife conservation. Over these years, Inogwabini has demonstrated leadership in strategic planning, program development, program management, team leadership and problem resolution through numerous programs he has led in the field across Central Africa. Inogwabini has published several dozens of peer-reviewed scientific papers, chapters of books and books. Inogwabini also owns and runs a third-generation farm whereby agro-ecology is practiced, combining traditional farming activities with forest regeneration and dynamics research, reforestation and carbon sequestration activities in addition to providing a recreational space to people. Inogwabini holds a PhD in Biodiversity Management (The Durrell Institute of Conservation and Ecology, The University of Kent, UK), an MSc in Conservation Biology (The Durrell Institute of Conservation and Ecology, The University of Kent, UK), an MA in Philosophy (Leeds University, UK) and a BSc in Physics (Université Pédagogique Nationale, Kinshasa).

Routledge Focus on Environment and Sustainability

Sustainability and the Philosophy of Science
Jeffry L. Ramsey

Food Cooperatives in Turkey
Building Alternative Food Networks
Özlem Öz and Zühre Aksoy

The Economics of Estuary Restoration in South Africa
Douglas J. Crookes

Urban Resilience and Climate Change in the MENA Region
Nuha Eltinay and Charles Egbu

Global Forest Visualization
From Green Marbles to Storyworlds
Lynda Olman and Birgit Schneider

Sustainable Marketing and the Circular Economy in Poland
Key Concepts and Strategies
Anita Proszowska, Ewa Prymon-Ryś, Anna Kondak, Aleksandra Wilk and Anna Dubel

Risk Management for Water Professionals
Technical, Psychological and Sociological Underpinnings
Anna Kosovac

For more information about this series, please visit: www.routledge.com/Routledge-Focus-on-Environment-and-Sustainability/book-series/RFES

Climate Perspectives from the Congo Basin

Bila-Isia Inogwabini

LONDON AND NEW YORK

First published 2024
by Routledge
4 Park Square, Milton Park, Abingdon, Oxon OX14 4RN

and by Routledge
605 Third Avenue, New York, NY 10158

Routledge is an imprint of the Taylor & Francis Group, an informa business

© 2024 Bila-Isia Inogwabini

The right of Bila-Isia Inogwabini to be identified as author of this work has been asserted in accordance with sections 77 and 78 of the Copyright, Designs and Patents Act 1988.

British Library Cataloguing-in-Publication Data
A catalogue record for this book is available from the British Library

ISBN: 978-1-032-79763-2 (hbk)
ISBN: 978-1-032-79765-6 (pbk)
ISBN: 978-1-003-49375-4 (ebk)

DOI: 10.4324/9781003493754

Typeset in Times New Roman
by KnowledgeWorks Global Ltd.

For Gracia, Galilee, Golden and Wivine
for being who each of you is for me.

Contents

Introduction   1

1   The Novelty and Immensity of the Climate Issue
and Social Inertia   13

2   International Good Political Intentions: Difficulties
to Implement Them in Central Africa   24

3   How Can Central Africa Successfully Play Its
Global Political Role in Managing CO_2 and Other GHGs?   39

4   Climate Change: Adaptation and Mitigation in
Central Africa and DRC   50

5   De-carbonating Developed Economies and the Right
to Development   63

6   Biodiversity Erosion, Climate Change and Life's Purpose   75

7   Capacities, Institutional Arrangements, Democracy
and Climate Change in Central Africa   96

Index   *109*

Introduction

The Earth Summit that was held in Rio de Janeiro in 1992 ignited an international movement on climate change whose most important and world-widely known event is the Conference of Parties (COP). Rounds of climate change's COPs started in Berlin (Germany) in 1995. Since then, people from all parts of the world meet annually to discuss how well and what needs to be done by all parties to stabilize or even reduce the levels of anthropogenic greenhouse gas (GHG) concentrations below the thresholds of the becoming dangerous for the earth's climate system. Keeping the human-induced GHGs below dangerous levels is the ultimate goal that the United Nations (UN) Framework Convention on Climate Change pursues.

Yet, despite being one event that is attended by most people from all parts of the world, the impression that one has from the end of each climate COP is that not so much progress is achieved. Every now and then, people complain about how little progress has been achieved in these mega celebratory events. This book identifies and discusses why climate change agenda does not move as it should in Africa, in the entirety of the countries of the Congo Basin and, particularly, in the Democratic Republic of Congo (DRC). Indeed, in Africa, as it is the case for the Congo Basin, and in the DRC, the impression one gets from the climate COPs is that they have, so far, had little tangible effect on the conditions of Africans and Congolese. This book strives to find reasons for this lethargy or lack of progress. Apart from the Africa-specific issues described below, the book argues that there are more fundamental reasons why climate actions have not taken off as rapidly as most people would have expected. These fundamental reasons, the book goes on to argue, are to be found in some general human attributes that make action to curb the levels of GHGs difficult. In taking this argument, the idea is not to argue for inaction or justify it, but the book argues that social inertia, latency of acquired modes of life and the very idea of trading in carbon credits explain, at least partly, why the meager observed concrete climate action does not match the huge aspirational wishes to see things and modes of lives changed to address the climate crisis.

The book is built on long discussions on climate change and how it is managed world-widely in Africa, the countries of the Congo Basin and the DRC

DOI: 10.4324/9781003493754-1

with students who participated in courses I teach. I have been having these heated discussions with students who have taken courses for their master's degree. The first of these courses was Climatology and Climate Change, which I taught to students enrolled in the professional master degree in agricultural and veterinary sciences at the Jesuit Loyola University of Congo. The second course is 'Cross Cutting Issues of Sustainable Development' that I convene for master's students in Sustainable Development at the School of Management of the Catholic University of Congo (SM-UCC) based in Kinshasa. The third course is that of 'Current Global Ecological and Environmental Challenges', which I convene for students enrolled in classes for their master's in Environment and Natural Resources Management at the SM-UCC. This very course is also given to students of the Francophone Senghor University (based in Alexandria, Egypt) and is attended by students from other parts of the continent. So, some ideas discussed in the chapters that follow also include views from some other parts of Africa.

While the course of Climatology and Climate Change is a more classic endeavor where the physics and chemistry of climate change are taught and the increases in the concentrations of GHGs in the atmosphere are explained as to how they make climate shift to unbearable levels, the courses of 'Cross Cutting Sustainable Development Issues' and 'Current Global Ecological and Environmental Challenges' have been of a different pedagogy and attracted audiences even beyond enrolled students. This has been so because these last courses provided a stall where students presented ecological problems they felt were among the most pressing, challenged commonly held views and discussed some paths toward mitigating these major issues. Obviously, stating that students were involved in the discussions that led to the writing of this book does not transfer the responsibilities of statements I make in the text that follows to the students. If anything, the responsibilities (if any) are entirely mine. However, recalling the participation of students in the progress that led to the writing up of this piece is important because not only did students help with checking ideas I presented as part of the courses but they actually contributed to reshaping some of the ideas I have held. In some cases, students had even shaken some of these ideas I held. From these discussions, it emerged that there are four Congo Basin-specific obstacles to making the expected progress to solve the climatic question. These main obstacles revolve around the notion of the right to development, the principle of common but differentiated responsibilities and the precautionary principle and the very format of climate negotiations. In addition to these reasons, the book shows that much of the failure to implement climate change resolutions in the Congo Basin is correlated with the lack of science that grounds its discourse (narrative) on African realities. To get out of it, the book injects to the global debate the idea of creating structures that can support research and the integration of the climate issue into the development process that is yet to dawn in Africa.

Meetings to debate climate change have become one of the defining moments of the international multi-lateral diplomacy over the last three decades. This is evidenced by the proliferation of protocols agreed upon and treaties signed since the UN1992 Conference on Sustainable Development. Since 1992, the world has gone from conference to conference in various formats and different levels (states, regions and world-widely). The aim of these meeting is the fundamental quest for a global human consciousness to prevent climate from reaching levels that life cannot support; the hopes invested in them are to reap a human consciousness to be used to globally address the climate question and the sustainability of life on the earth. Within this mix of various formats of consultations at all decision-making levels, COPs emerge as the most emblematic forum. Apart from the rather philosophical idea of raising human consciousness globally to address the climate question, lay people, in their majority in Central Africa, expect concrete solutions to combat the effects of climate change and stop the furtherance of deteriorating climatic conditions. It is in line with this search for concrete solutions that it is quite natural that each COP gives birth to resolutions intended to bring the earth back to a climatic trajectory that constantly bears life. One can, doubtlessly, qualify COPs as making an essential part of the process to politically manage carbon dioxide (CO_2) and other GHGs. This is so even though the strongest impression remains that the resolutions taken during climate COPs are becoming less and less practical. For ordinary people of Central Africa, resolutions are less applicable in the concrete daily life of humans.

Given the impression of lack of tangible progress from these numerous and large meetings, the fundamental question that springs to the minds of many of the students has been why so many meetings, so much noise and so many resolutions but little practical progress? This book discusses possible reasons why the execution of the resolutions resulting from the climate COPs is very slow in Africa. From this standpoint, the book does not go through resolutions of all COPs one by one. Resolutions resulting from each COP are very numerous, and reviewing them would be a daunting task even as the exercise remains plausibly and materially feasible. The ambition of this book is even less than the one of reviewing the policies of each Central African states. The book focuses on the major issues that make resolutions and policies on climate change difficult to apply today nationally (DRC), regionally (in Central Africa) and internationally.

Stepping back to the origins of the book and how it was shaped, it is worth recalling that all the students who I interacted with during the courses described above felt, in one way or another, that climate change was the most important current global ecological challenge the world currently faces. These discussions narrowed down to the question how climate change challenges Central Africa. With concrete examples, students demonstrated that Central Africa was most affected by the effects of climate change; it also encountered many challenges to garner the motion to address the challenges climate change brought forth. While students acknowledged that challenges Central Africa faces are both identical to those encountered elsewhere in the world, it was also clear that some other challenges are particular to the regional culture, geography and political history. Clearly, discussions with students and other participants to my courses made it plain that challenges faced by ordinary citizens of Central Africa are direr. This was felt to be so because internal social structures and historical facts that brought Africa where it currently stands also caused the direness of climatic conditions, at least partially.

How Central Africa will feed its impressively growing populations in the decades to come was the first situation most people felt to worsen because of climate change and leading to even direr humanitarian, political and social conditions. Difficulties to address challenges in food production were thought to be historically built-up but also to become more vexing with changing climate conditions. My argument with the students was that Central Africa is fully capable of producing sufficient food to feed its populations, even under changing climate conditions. This argument was nuanced because this possibility would be translated into a tangible action only if ecological factors were taken into account in regional political planning processes. Sound political planning processes incorporate ecological evidence-based data and should include dividing Central Africa by its natural agricultural functions and create schemes of internal regional food trade. However, the 1885 borders remain so strongly felt both politically and emotionally throughout Central Africa to the point that collective regional efforts to address such difficult questions such as hunger will take enormous labors to overcome. While the effects of climate change are reducing the size of cultivatable lands across the region, internal increased human populations are also changing the geographic space occupancy patterns. Geographic occupancy also changes because of the cumulative effects of the presence of large industrial ventures in mining, logging and agriculture. The modern African state's nature and the sovereignty notion strongly tied to it impede the possibilities to cross regional borders. Following historical paths to get to where suitable climatic conditions prevail in the climate change era is politically difficult.

The global inaction of the world's most developed nations in the face of climate-induced calamities was equally hotly debated. Responses to the question varied in their tones and originality. Most people who were part of these discussions rightly pointed out that the main reasons this was so were a lack of political will, bad public policies and willingness to keep historically acquired social advantages in developed countries. For people in Central Africa, the willingness to keep and defend historically acquired social advantages by most developed countries transpired in the fact that most Western politicians were reluctant to take concrete action to protect themselves against the possibilities that climate-sensitive policies would lead their electorates to eject them from the power. For most people in Central Africa, in technologically most advanced nations, concrete climate change action equates to living decently without being jammed with useless gadgets. Living decently means to eliminate the most eccentric desires that unnecessarily torment daily lives and travel slower. Getting the climate actions implemented would, politically, amount to disconnecting the majority of people in developed countries of the current life standards. Disconnecting people of the hardly acquired high life standards is the hardest thing to get accepted. Technologically developed communities have become psychologically drugged by instantaneity and speed. As an example, the attempt to introduce an ecological transition tax in France aroused fiery popular outcries. This confirms how difficult it is to get rid of acquired habits. Altering acquired comfortable life is the hardest thing to get accepted even by those who wish to see concrete climate change action. The claims of 'Yellow Vests'[1] were legitimate because the ecological transition tax, if it was implemented, would have hit the most fragile of capitalism. However, paying the price of the de-carbonation of the economy remains necessary to save the earth from disasters that climate change is to cause if humans do not manage to limit its effects.

The inaction to curb climate change is also general across the world and cultures. This suggests that inaction on climate change may also be due to a more profound reason, which is that human history lacks a precedent where a situation of the magnitude of climate change had faced humanity. This goes beyond a lack of political will, bad public policies and willingness to keep and defend acquired social advantages. Humans have no recorded memories of climate hitting the environments where they resided and wiping out the biodiversity and other means they could use to sustain their livelihood.

Climate change is a geologically historical and natural phenomenon[2] even in its modern formulation. The Swedish Physicist Svante Arrhenius described human-induced climate change for the first time in 1896.[3] Naturally, solar rays entering earth atmosphere are partly absorbed, whereas another part is reflected back to the outer space.[4] Arrhenius showed that GHGs released into the atmosphere formed an opaque layer that prevented infrared rays from escaping back into the space. Ensnared within the earth atmosphere, infrared rays heat the earth more than would be the case under conditions where they would escape back to the outer space.[5]

Summarily, scientists agree that the climate has always alternated between episodes of much colder temperatures and those of warmer ones during the Holocene. Holocene is the geological period that began since 150,000 BP and runs until our own time. Recently, some researchers[6,7] introduced the Anthropocene era, which is to mark the end Holocene and start a period where humans dictate the course of things on the earth.[8] Anthropocene rises debates some[9] arguing that the concept is ideologically driven. Whether we are still in the Holocene or we actually entered into Anthropocene, industrialization has played a key role in the last two centuries; it continues and will continue to significantly influence the patterns of shifting climate if humans are not engaged in changing the current course of things. Indeed, throughout the Holocene, the earth experienced four episodes of cold and arid climate (glaciations) and four hot and humid periods. Glaciations occurred between 150,000 and 130,000 BP, 115,000 and 70,000 BP, 22,000 and 13,000 BP and 11,000 and 8,200 BP. Warmer and wetter periods occurred between 130,000 and 115,000 BP, 70,000 and 22,000 BP, 13,000 and 11,000 BP and 8,200 to the present.[2] These dates are not as clearly edged as they seem at first sight. There have always been interglacial periods[8] (transition times between geological periods).

For Central Africa, studies[10,11,12] show that forests retracted during glaciations while they recolonized the lost ground once humid and warm conditions returned. During glaciations, researchers assert that Central African forests were less luxuriant compared to what they are today; they were reduced to congruent portions around the mountains and along large rivers. These patchily spread forests around the mountains and along large rivers were the *refugia* from which current forests sprang up to recolonize the territorial extent they cover today.[13] From this long historical perspective, one logically poses the question why humanity worries so much about climate change. This question is answered by carefully looking at the evolution of GHGs over the last millennia, as detailed in the continual reports of the Intergovernmental Panel on Climate Change (IPCC[14]) produces. IPCC reports show that the concentration of GHGs (CO_2, methane (CH_4) and nitrogen dioxide (N_2O)) remained sensibly identical over millennia but suddenly substantially changed by 1880. From 1880 onward, the concentrations of CO_2, CH_4 and N_2O in the atmosphere continually peak upward.[15] The year 1880 represents the moment when the industrial revolution of the Western world began, coinciding with the invention of the steam engine, the use of coal technology, the mechanization of cotton weaving and the mastery metallurgy in England.[16] Industrialization releases GHGs in quantities that are far greater than natural earth's emissions. The energy needed to keep up with the industrial civilization emits many GHGs into the atmosphere; many GHGs ensnare more infrared rays and cause abnormal levels of heating. Why do humans speak more of climate change now than before? The concern is that heating levels are nearing the thresholds that life can no longer naturally bear. Human actions increase heat.

This book is premised on the fact that human influence on the earth climate is factually established.[17] Anthropogenic GHG emissions continue to increase, despite meetings and debates over climate change. Anthropogenic GHGs emissions come from industrialization and the need for energy to keep human businesses running.[18] These premises lay the background of the discussions that pervade all the discussions on climate change. Most people now know what caused the climate to change and what needs to be done to shift the current climate trends back to normal. Few people, however, feel it is their responsibilities to act for change. With the background that climate change results from industrialization and past development actions, climate discussions evolve essentially around the notion of responsibility and justice. For Africans, because industrialization caused climate to change, industrialized nations should bear the costs of fixing it. This is why Africans strongly grip on the notions of the right to development, principle of common but differentiated responsibilities and the principle of the polluter pays. Africans insist that the Loss and Damage Mechanism, which was introduced by the 2013 COP19 (Warsaw, Poland), should be implemented. The Loss and Damage Mechanism recognizes that certain unavoidable impacts of climate change are consequences of GHGs emitted by Western developed countries that now cause sufferings in the Global South. Unfortunately, the losses and damages triggered by the effects of climate change are unevenly scattered across the planet. As such, the Loss and Damage Mechanism is a matter of climate justice.[19] Pressed to take action, Africans stand by the will to develop their own countries even if developing their countries would imply emitting more GHGs. Latin America, the Arab-Persian Gulf Monarchies, Russia and African countries that produce crude oil, which built their economic prosperity heavily depending on trading fossil fuels, take the suggestion to de-carbonize the industry as a political trick to impede their economic prosperity and decrease their growing global influence; they are far from giving up the prosperity they built and the influence they are gaining globally through the trade of fossil fuels. Countries like China and India whose economic prowess of the last three to four decades is hugely carbonated also think decarbonizing economies would spell the doom of their own economic prosperity; they think it would be unjust to ask them to phase out of carbonated economy under current conditions. Even in the most prosperous Western countries where industrialization emitted more GHGs than any other parts of the world, some outstanding voices suggest that to demand punitive[9] actions to change the current trajectory of the earth's climate is not necessary; actions need more reflection, and the attribution[20] questions need to be clarified before climate actions can be endorsed by all.

The content of Chapter 1 is about the fact that humanity has no recorded memories on the climate change. It explains why politicians have been the least sanguine to take bold action to deal with current climate change consequences. The chapter's main argument is that climate change being a relatively novel and immense situation, social inertia prevents us from clicking on the emergency psychological chain reactions. Emergency psychological chain reactions would have compelled people to act. Despite the lack of historic precedents, the recent Covid-19 pandemic has shown us that when pressed by the needs to stay alive, humanity can act efficiently. This should give us hope for the future. With the emergency chain reactions not yet ignited, questions arise on if there are possibilities for Central Africa to begin dealing with difficult situations (made of losses and damages) that its communities already face because of climate change.

Chapter 2 discusses the question of the possibilities for Central Africa to begin dealing with difficult climate situations. Rights of communities of Central Africa to develop materially and achieve their well-being constitute the main obstacle to tackle the difficult issues of climate change. Most people in Central Africa hold the view that climate change should not be used to prevent them from working toward their own economic and material development. The right to material development and wellbeing is in the Rio de Janeiro convention. It should address this sticking point. Equally, most people in Central Africa feel that the principle of common but differentiated responsibilities does not clearly help situate the responsibilities of the world's most industrialized nations in the climate crisis. This claim becomes hotly debated and constitutes probably the most important sticking point because most financial promises to offset historical climate responsibilities have not materialized long years after they were made. Within this perspective, even the precautionary principle becomes a hoax used by developed nations to refrain non-developed countries under the yoke of poverty. People take this to be plausible even though when pressed to think carefully, they agree that the precautionary principle is, in its essence, politically neutral. For communities in Central Africa, demands for them to bear their fair share of the climate change's cost are unfair when promises of those who polluted more in the past have not been kept. To keep these promises, fresh funds are to be made available. A new insight in Chapter 2 is that the people of Central Africa think that the very format of climate negotiations is doomed to lead to failure. Climate change negotiations are arguably done in formats that are far too remote from capacities of non-trained diplomats from Central Africa to grasp. Furthermore, the jargons used in the COPs reflect the same international political order where the most powerful nations have the last words in the global communiqués that are published at the end of each COP. With such a relatively dim perspective, is there any way around these obstacles?

Chapter 3 identifies a lack of appropriate climate change structures as one of the major obstacles that countries of Central Africa to confront climate change. Lacking such structures, governmental actions to fight climate change cannot be coherent enough to instill impactful changes. For the DRC (Central Africa) to successfully play its global political role in managing the GHGs, it should create structures whose mission is to support research and create climate knowledge grounded in the local context. Creating structures to support research and climate education in Central Africa should become part of the programs and processes to sustainably develop Central Africa. Funding to address climate change and its effects is the most difficult challenge governments of Central Africa face; to fund the search and creation of new knowledge, the DRC and Central Africa have to rely on internal financial means to support their own efforts. These resources can be mobilized by integrating climate change in the scheme and process of sustainable development.

Chapter 4 takes on the question that given that climate negotiations are the ground where the economically and politically stronger states deploy their might, is there any way around these obstacles? One sensible way out of this catch-22 situation is presently imbedded in the ample possibilities that Central African countries have to contextualize the concept of sustainability to match their own needs and implement whatever is climate-sensible for their own situation. To do this, the people of Central Africa should not try to reinvent the wheel; they should use the globally accepted climate change action plan but contextualize this for their own conditions. But everyone should participate, within the limits of their capacities, to maintain the social safety net. Everyone should support the upward social progress of all, and there should be efforts to re-enchant the life in traditional and remote villages. Re-enchanting life in rural areas should be packaged with a clear national land use planning. The land use planning process is rethought as a strategy for both adaptation to and mitigation of climate change. Above all, Central Africa should break away from the propensity to mimic, without proper examination, anything that is American or European. Central Africa and the DRC should avoid the temptations to think that there is only one development model, which is epitomized by currently most advanced countries. Re-enchanting life in rural areas should also mean to build more middle size cities to release the pressure the growing populations would exert on habitats and biodiversity globally.

Chapter 5 discusses the de-carbonation of economies in developed countries, the emerging ones and the technologically retarded nations. De-carbonating economy is a big topic and a more controversial issue across Central Africa. The citizens of Central Africa (Congolese) hotly debate de-carbonating economy because most people suggest that the whole issue of de-carbonizing economic development is an attempt by developed countries to prevent less developed nations from progress toward sustainable development. The argument goes that de-carbonating economies will negatively impact development perspectives for countries in Central Africa. Chapter 5 conveys the often-heard argument that it is far from being morally plausible to ask countries in Central Africa to limit their development ambitions to support the continued opulence in the developed world. Legitimate ambitions for development the people of the Congo Basin dearly cherish should not be left unimplemented because of the priority to be given to the climate crisis. People across the Congo Basin always ask the question why what developed nations did in the past cannot be done in other contexts. In this vein, ecological neo-colonialism and ecological imperialism are often brought forward. While having development ambitions is a legitimate and sane way of thinking, development should not be viewed as the function of the carbon singly; there might be and there are alternatives that may even be ecologically less costly. The main issue is to identify these alternatives and use them instead of relying on carbonated economy. Carbonated development would, in the end, wash away all the efforts through extremes weather conditions that increased GHG concentrations in the atmosphere will cause. It is not because Western industrialization was a very carbonated process that people from Central

Africa should do the same thing. It is not because something was done in the past that is something good to be repeated. One should not repeat mistakes of the past simply because they were made. This recalls the ideas discussed in Chapter 4 that Central African countries need to contextualize the concept of sustainability to espouse their cultural, ecological, environmental and historical conditions. By embedding development actions in their context, they can find out climatically and ecologically less onerous transformational technologies and process to sensibly progress and achieve sustainable development goals. Contextualizing global solutions is a pathway that avoids repeating old development ideologies, which led the world to the current global climate crisis. Old development ideologies were framed around the idea to copy whatever has been done in the Western world; they are not necessarily good for Central Africa's conditions.

Chapter 6 presents the more complex and vexing question about what are the meanings, if any, of all the efforts being invested in tackling climate change? The people of Central Africa often ask the question about why is there the drive to invest so much time, energy and funds in searching solutions to the climate crisis? Chapter 6 takes up this question and broadens it to include also the global erosion of biodiversity. Both climate change and biodiversity erosion crises are linked and pose a deeper question of the purpose of life itself. Whatever people take as meaning of life (teleological, biophilia and species-intrinsic values) dealing with both crises is very important because both climate change and the erosion of biodiversity threaten the very possibility of there being life on the earth. Arguably, there is nothing worthier than life itself. Efforts that humans invest in reducing the concentrations of GHGs in the atmosphere and the ones devoted to keeping ecosystems, species and biological functions thriving are the best thing one would do for the people one loves and to oneself. People owe it to the ones they love, their lovers and to themselves.

Chapter 7 discusses what possibilities are there for countries in Central Africa to address their own communities' needs vis-à-vis the climate and biodiversity crises. I pledge that the current institutional capacities and arrangements in Central Africa are inadequate to cope with both crises. The lack of effective political participation due to the long-time delayed exercise of true democracy in Central Africa makes it difficult for questions of climate change to find satisfactory solutions. Unfortunately, this fact will remain difficult in the long run. If there are any supports that can help these countries find solutions to the climate and biodiversity crises, it should be in the aid to create truly functional and functioning democracy whereby views and voices of local communities would be heard and accounted for in the making of public policies. Truly functional and functioning democracy does not mean to mimic everything from Western industrialized countries; the concept itself should be contextualized and anchored on the culture, ecology and history of the region.

· I discussed some of the content of Chapter 6 with Professor Nathalie Frogneux of the Catholic University of Louvain La Neuve, Belgium. The chapter comes from a proposal I discussed with her for a possible PhD dissertation in Philosophy, which would have looked at the question of meanings in the middle of the double biodiversity and climate change crises. Nathalie was keen to work together on this interesting project, but the administration of the Catholic University of Louvain La Neuve subjected me to conditions to I felt were unjust, which prevented me from pursuing that opportunity. Despite these discussions, Chapter 6 is entirely mine, and the entirety of Nathalie's comments were left out because I wished to make this work and its contents a total responsibility of mine. So, any flawed thinking and mistakes that one would find in this book are singly mine and should not be blamed on anyone I have interacted with during the preparation of the manuscript. I can only hope that there are not many mistakes and that flawed reflections have been kept at minimal levels. The final version of the book has been improved by comments I received from two anonymous reviewers to whom I owe huge thanks. At Routledge, I benefited from the support of Annabelle Harris whose patience with me and guidance were so critical to make the book project become a reality.

Notes

1 Bate F (2018) France's Macron learns the hard way: Green taxes carry political risks. Reuters 2 December 2018. Edouard Philippe, then premier minister of France, introduced a 16% tax on both fuel and petrol in 2018, with the goal of contributing toward cutting France's CO_2 emissions by 40% by 2030. Philippe believed the new increase in carbon tax would push the fuel and petrol's users to change behavior and protect the environment while positively boosting the use of cleaner energies. Despite being overtly pro-climate, the lower socioeconomic classes felt that they were bearing the largest burden of the CO_2 tax while carbon-intensive companies were exempted. This interpretation led to social movements and riots across France.

2 White LJT (2001) The African rain forest: Climate and vegetation. In Weber W, White LJT, Vedder A, Naughton-Treves L (Editors) African Rain Forest: Ecology and Conservation. Yale University Press, pp. 3–29.

3 Arrhenius S (1896) On the influence of carbonic acid in the air upon the temperature of the ground, in the London, Edinburgh and Dublin. Philosophical Magazine and Journal of Science XXXXI (5): 237–276.

4 Maslin M (2004) Global warming: A Very Short Introduction. Oxford University Press, p. 162.

5 Rohli RV, Vega AJ (2017) Climatology (Third Edition). Jones & Bartlett, p. 443.

6 Poff NL, Matthews JH (2013) Environmental flows in the Anthropocence: Past progress and future prospects. Current Opinion in Environmental Sustainability 5 (6): 667–675.

7 Lewis SL, Maslin MA (2018) The Human Planet: How We Created the Anthropocene. Pelican, p. 256.

8 Ellis E C (2018) Anthropocene: A Very Short Introduction. Oxford University Press, p. 183.

9 Ferry L (2021) Les Sept Ecologies : Pour une alternative au catastrophisme antimoderne. Editions de l'Observatoire/Humensis, p. 274.

10 Maley J (1990) Synthèse sur le Domaine forestier africain au Quaternaire récent. In Raymond LANFRANCHI & Dominique SCHWARTZ (Editeurs), Paysages Quaternaires de l'Afrique centrale atlantique, Paris, Office de Recherche Scientifique et Technique d'Outre-mer (ORSTOM), pp. 383–389.

11 Maley J (1991) The African rain forest vegetation and palaeoenvironments during late quaternary. Climate Change 19: 79–-98.

12 Maley J (1996) The African rainforest: Main characteristic of changes in the vegetation and the climate from upper Crestaceous to Quartenary. Proceedings of the Royal Society of Edinburgh 104B: 31–73.

13 Maley J (1996) Fluctuations majeures de la forêt dense humide africaine au cours de vingt derniers millénaire. In Hladik CM, Hladik A, Pagezy H, Linares OF, Georgius JA, Koppert, Froment A (Editeurs) L'alimentation en forêt tropicale: Interactions bioculturelles et perspectives de développement, Paris, UNESCO/CNRS/ORSTOM, pp. 55–76.

14 Intergovernmental Panel on Climate Change (IPCC) Climate Change (2014) Summary Synthesis Report for Policymakers, Geneva, United Nations, p. 33.

15 Maslin M (2013) Climate: A Very Short Introduction. Oxford University Press, p. 159.

16 Northcott MS (2007) A Moral Climate: The Ethics of Global Warming. Dalton, Longman and Todd LTD and Christian Aid, p. 336.

17 Wijkman A, Rockström J (2012) Bankrupting nature: Denying Our Planetary Boundaries (revised edition). Earthscan/Routledge, p. 206.

18 Church of Sweden (2008) Uppsala Interfaith Climate Manifesto 2008. Church of Sweden, p. 115.

19 Mechler R, Calliari E, Bouwer LM, Schinko T, Surminski S, Linnerooth-Bayer JA, Aerts J, Botzen W, Boyd E, Deckard ND, Fuglestvedt JS, González-Eguino M, Haasnoot M, Handmer J, Haque M, Heslin A, Hochrainer-Stigler S, Huggel C, Huq S, James R, Jones RG, Juhola S, Keating A, Kienberger S, Kreft S, Kuik O, Landauer M, Laurien F, Lawrence J, Lopez A, Liu W, Magnuszewski P, Markandya A, Mayer B, McCallum I, McQuistan C, Meyer L, Mintz-Woo K, Montero-Colbert A, Mysiak J, Nalau J, Noy I, Oakes R, Otto FEL, Pervin M, Roberts E, Schäfer L, Scussolini P, Serdeczny O, de Sherbinin A, Simlinger F, Sitati A, Sultana S, Young HR, van der Geest K, van den Homberg M, Wallimann-Helmer I, Warner K, Zommers Z (2019) Science for loss and damage. Findings and propositions. In Mechler R, Bouwer LM, Schinko T, Surminski S, Linnerooth-Bayer JA (Editors) Loss and Damage from Climate Change: Concepts, Methods and Policy Options. Springer, pp. 3–36.

20 James RA, Jones RG, Boyd E, Young HR, Otto FEL, Huggel C, Fuglestvedt JS (2019) Attribution: How is it relevant for loss and damage policy and practice? In Mechler R, Bouwer LM, Schinko T, Surminski S, Linnerooth-Bayer JA (Editors) Loss and Damage from Climate Change: Concepts, Methods and Policy Options. Springer, pp. 113–153.

1 The Novelty and Immensity of the Climate Issue and Social Inertia

Introduction

When one is asked the question why the resolutions taken at the climate change COPs don't have a grip on the real lives of citizens throughout the world yet, the first reason that should come to mind is that climate change, in its current causes and dimensions, remains a new fact. Admittedly, global climate change is a natural phenomenon that happened constantly throughout the geological history of the earth. Historically, in a geological timeframe, the earth's climate alternates between warm and cold periods.[1] Similarly, it is a fact that the current climate change began its unprecedented magnitude in the 1800s; this coincides with the first industrial revolution. However, it remains equally true that the public awareness of the extent and consequences of the current climate change dates from the 1970s when the Report to the Club of Rome was published.[2] That awareness remained within very small circles of the world's scientific elite. Chronologically, the emergence of this collective consciousness on climate change takes its current shape only from the 1992 Rio de Janeiro Conference.[3] Indeed, 2022 marked three decades since this collective climate consciousness has been developing with an increased magnitude. It was during the same three-decade period that research was deployed to find adequate climate solutions. Three decades may seem long enough for a human life, but much more time is needed to initiate the cultural, technological and civilizational changes needed to come to grips with an extraordinary issue of human civilization. Climate change took long enough to reach its current level of maturation. At the modern civilization history scale, the climatic question remains a novelty in its current amplitudes. Thus, our climate knowledge, well certified by science, will still take time before becoming practical actions. What the world currently needs to overcome the climate crisis is a total upheaval of human civilizational paradigms. This shift in the type of the life people have been living for centuries needs to involve structural readjustments and changes in societal models. Viewed from this perspective, the resolution of climate change questions is not a simple matter of days, months or years! Political, social and cultural action will take a long

DOI: 10.4324/9781003493754-2

time. This should not be understood as a call to waste more time before starting to put in place the necessary measures! On the contrary, it is a necessary perspective so that the people of Central African are not disappointed by the little progress recorded to date and over the coming decades.

Social Inertia and Resistance against the Novelty as Part of the Climate Problem

Why should the cultural, political and social action to address climate change take long time before their actual effects can be felt? At least partially, the answer is because of social inertia.[4] Social inertia is the natural resistance against novelty. Social inertia is the time it takes to overcome acquired social and cultural attitudes. This equates to stating that the psychological dissipation time required to positively react to negative signals emitted by the effects of climate change is taking and will take longer than it is currently expected when the international community meets every year in the COPs to discuss what is to be done to bring human-induced climate change under controllable scales. Psychologically, the dissipation time is the period it takes for individuals to respond to a provoking situation.[5,6] For the socially acquired behavior, the dissipation time is simply the time it takes to dissolve and relinquish the already absorbed social and cultural norms. It is the time necessary for tossing up deeply ingrained traits of collective behavior.

The knowledge of climate change and its associated present and projected effects are such shock-provoking events that discussions on climate change during the COPs aim at igniting individual and collective action to stop extreme disasters from happening. One expects immediate reactions, but the pattern, so far, is the one where there is a long delay in responding to the danger's red signals that humanity faces if climate change is not squarely and timely addressed. That is so because socially acquired behaviors become cultural norms; they have been distilled and absorbed over a long time. To dissipate, dissolve and relinquish these slowly built social and cultural norms mean to throw up deeply ingrained traits of collective identity. Psychologically, culturally and socially,[7] it is hardly surprising that there is a delay between the inducement of the COPs' climate change message and the responses from people across the world.

Arguably, that there is a delay between the stimulus and the response even when there are opportunities to take action against climate change is somewhat expectable. Expecting things to go as smoothly as in a dream would be proof of naivety. As it is of public knowledge, individuals have bounded capacities to perceive, recall, interpret and calculate even things that are in favor of their own self-interest. Individuals often have just pieces of information to make decisions. Even when individuals have sufficient information, each type of information will carry different weights for non-logical and time constraint reasons.[8] The time to react to a situation and change in the attitude of individuals varies according to different individuals; it is logically expectable that the total dissipation time of habits acquired by communities over many centuries takes a long time. At least part of the communities whose leaders come to the climate change COPs arrive in these meetings bearing with them values, norms and social conducts engrossed by long time frustration-instigated attitudes.[9] Surely, it is worth recognizing the fact that climate change is widely acknowledged as an outcome of the development of most Western countries. That Western countries are the most developed nations frustrates most under-developed countries to be demanded to bear the costs of repairing the climate damages. From this perspective, it is explainable that the countries of these less well-off parts of the world sometimes exhibit climate insensitive behaviors. It is harder for them to accept and move on[10] with the proposed positive collective action. This comes in addition to the fact that the habits that humans have acquired so progressively took millennia before they reached the current situation. In many ways, it seems rather overoptimistic to expect these habits to be changed overnight. Even if people are aware of the perversity of a situation in which they find themselves in, getting rid of the reflexes, habits and lifestyles built up during the humanity's social evolution will only be gradual and will take entire generations. Undoing cultures is not an easy task, and, unfortunately, this applies to actions to tackle climate change too. Unfortunately, it could not be thought of otherwise because cultures frame how individuals perceive and explain their environments. Cultures also frame how sensitive people exposed to environmental changes can behave. Cultures frame how people experience exposure to dire environmental conditions is dealt with. Cultures form and inform the perceptions of risk[11]; they also affect the adaptive capacity of those exposed ecological and environmental risks. By playing such a significant role in shaping feelings and responding to environmental crises, cultures outline ways in which different people are related to and weigh equity and environmental justice.

Social inertia and resistance narrative may seem to be very pessimistic, and understandably so, against enormous efforts humanity spends to get the climate change right. However, taking the idea of the psychological dissipation time into account for socially and comfortably acquired positions and situations and accepting the fact that social inertia produces intolerance to changes[6] and it may not be wise to expect abrupt, profound and global metaphysical transformation immediately.[12] Of course, public indoctrination could seek to address social inertia and generate a momentum that would lead to collective actions to shift things around. In many ways, the COPs in their current formats are partly a process of indoctrinating the global community to take climate change seriously. However, even indoctrination needs substantial time to become a routine. Substantial indoctrination to achieve the much-sought positive climate action will require long-time mental adjustments.[13] Along this process, some of these mental adjustments may even be maladjustments, as it is often the case with massive socioeconomic processes. Not only does climate change require long periods for recovery and return to normal conditions but it also demands new points of ecological balance to be reached.[6] The planet the climate change heats up, biodiversity losses it causes and societies and cultures it reshapes also require significantly long time spans to be accepted by humans for them to adapt to the new normal conditions imposed by climate changes. It will require a long time and efforts for the human brains to adapt to this new normal.

Evolutionarily, humans have evolved as hunter-gatherers, living in groups where individuals had established roles and lives; though this is sometimes dangerous, life premised in these conditions was largely predictable. The brain evolved to be remarkably good at recognizing patterns and building habits, turning very complex sets of behaviors into something we can do on autopilot. The human brain evolved to be uncertainty-averse.[14] When things become less predictable, humans experience a strong state of threat. Faced with doubts, the chain of reactive consequences that these uncertainties generate in humans is most important. From a theoretical point of view, it is logical to infer that the anxiety uncertain situations can bring would lead to three different directions: (1) humans may feel forced to fight to retrieve their ancient conditions or adapt to the novel conditions, (2) they may feel threatened and distressed if and when the uncertainty is so strong[15] that humans may feel doomed or simply incapable of facing the new conditions or (3) they may simply ignore the threat and continue with their old normalcy.

That climate change is an area where the uncertainties of life abound is rather a truism. The discussions internationally held during the COPs on climate change and other similar meetings are to craft reactions that would follow the first of the three reactions above. The COPs push for actions for humans to feel compelled to fight to retrieve their ancient conditions or adapt to the novel conditions. So far, the observed reactions fall mostly within the last two theoretical responses above; in less developed countries, particularly in Central Africa, most people feel seriously threatened and distressed because the uncertainties imposed by climate change are so strong that people doomed and simply incapable of bearing the new conditions. What is felt from the most industrialized world is that, while some groups are a lot more vocal than the rest of the world, some significant portions of populations simply ignore the threat and continue with their time-worn normalcy.

The current observations after all the discussion around the plight that humanity faces due to the present and future consequences of the changing climate seem to indicate that most humans acknowledge that something totally different is pointing out in the horizons. In Africa, climate change-induced deviations in weather patterns bring communities to feel that something is shifting in environments where they live. For a long while now, studies[16,17] make it clear that climate change affects lives of African communities. With that reality, it would be expected that the majority of people would have felt compelled to take strong action to enable communities to retrieve ancient conditions. Advocates of the mitigation to climate change promote this path. For the mitigation to succeed, it is thought that humans who emit greenhouse gases (GHGs) can reverse these changes by adopting attitudes and activities that reduce GHG emissions. The thinking in the background is that reducing GHG emissions will hold because the current climate change crisis is human caused. That is why mitigation is, in its mild and non-high-tech versions, a process to reduce the sources or enhance the sinks of GHGs.[18] The non-high-tech version principally focuses on technologies that would enable the reduction of fossil-based CO_2 emissions.[19] Unfortunately, it seems that actions taken so far are below the thresholds to bring climate back to normalcy. The Congo Basin is the least prepared; it is the most vulnerable[20] among the vulnerable parts of the world.

That Africa and the Congo Basin are the least prepared and the most vulnerable to effects of climate change doesn't mean that things are not worrisome for other parts of the world. Climate change affects the earth globally. Disaster types and the magnitudes of climate events when and where they occur are the changes in climate change effects per geographic location. The level of civil and social preparedness also differentiates societies. People keep using fossil-based CO_2-emitting technologies throughout the world; long lived life's routines continue as if there is nothing happening that will globally affect everyone everywhere. Things look like if the majority of people across the world have chosen to ignore the threat climate change poses on their daily lives and choose to continue with their old normalcy. This is so despite the use of powerful media tools, high-level political discussions and numerous agreements. It is rather ironic and questionable if the discussions are not simply so high that the lay citizens are unable to grasp what is the motivation behind all the noises that come out of climate COPs. Irony set aside, it seems that the majority of humans have not been taking the first expected path whereby they would have felt forced to fight to retrieve ancient conditions. The belief that humans will adapt to the novel climatic conditions and continue to thrive as in the past can partially explain this general attitude.

The Immensity of Climate Change Consequences Is Not Recorded in the Human Collective Memory

The immensity of the projected climate change consequences doesn't appear to be something that individual and collective memories can recall. Climates have been changing naturally over geological times. The geological time scales are longer than any human culture. Phenomena that repeat over geological time scales are difficult to remember by humans who live three to four generations. The immensity of climate change consequences and lack of collective memories of such big events render improbable the idea of taking stringent measures against climate change. To sort out the mess humans created with climate, tough measures are needed to shift entire human cultures and their modes of life; they will touch the material conditions that founded the industrial and post-industrial civilization. There is no precedent recorded in the human collective memories from which to draw lessons to be learned.

The irruption of the Covid-19 pandemic in late 2019 gives a counter-example of the lack of a precedent argument that goes that the immensity of the felt consequences of an event and the lack of its historicity render the action to squarely deal with that event improbable. The lack of historicity should be understood that an event had never happened throughout the past. However, in the case of climate change, the event has been occurring in a cycle that goes beyond human history. If anything can be learned from the irruption of the Covid-19 pandemic, it is that the Covid-19 crisis showed that humans are capable of taking very strong actions when their lives are threatened. Shortly after the World Health Organization (WHO) declared Covid-19 a global pandemic[21] (11 March 2020), world's governments sternly took strong measures to stop the spread of the virus, and countries closed airports. The movements of people and goods were stopped across the globe; critically needed movements were rigorously controlled. The world health authorities made the mask wearing mandatory for people attending events in closed areas and in most cases even in open space. Work offices were closed, and work from home became mandatory.[22] In African countries, some ceremonies with cultural and social meanings such as funeral ceremonies were changed to accommodate social distancing policies to limit the geographic and social spread of Covid-19.[23] Countries adopted war-time rhetoric through declaring a health state of emergency; weekly government ministerial cabinet meetings were held regularly and provided daily and weekly public debriefs on the evolution of the contamination and death tolls caused by Covid-19.[24] Wealthy nations' governments invested significant funds to help laboratories to develop a vaccine.[25] This latter action enabled most countries to resume the normal public life's activities in a relatively short time span.

But Humans Can Take Necessary Measures When Their Very Existence Is Threatened

That the above measures were taken in a short time after the WHO declared Covid-19 to be a general threat indicates that humans are capable of taking necessary measures when they face existence-threatening dire situations. This also raises the question why climate change, with the projected catastrophic consequences it is likely to bring to the world, is not receiving similar attention. Sensibly, the public response to Covid-19 diverges from the reaction that the same world public reserves to climate change because Covid-19's hazard was felt immediate, whereas climate change effects are supposed to be non-immediate. The degrees of felt urgency explain the difference between the strong human anti-Covid-19 reaction and the feeble reactions to climate change. At the climax of the Covid-19 crisis, the death tolls in higher income countries were so staggering that there could be rapid and massive reaction. Officially, Covid-19 most severely hit the United States, India, the United Kingdom, Brazil, France and Germany. Faced with these high death records, public opinions in these countries demanded strong measures. In liberal democracies, public opinions weigh heavily on processes to access and exercise political power; the reactions from the governments of these countries could only be sufficiently strong. Apart from the pressure from public opinions in the wealthiest countries, two other related factors were combined to provide substance to this globally strong reaction in France, Germany, Italy, the United Kingdom and the United States. The first factor was that these countries are the financially wealthiest world nations and could invest massively in research to find a rapid technological solution. The second factor was that these were technologically the most equipped countries. The combination of wealth and technology made investments to inflect down and stop the death tolls successfully in a short period of time. Research conducted to find vaccines were commensurate to countries' development levels. The speed at which the wealthiest world's nations strongly reacted against Covid-19 was so rapid because these nations were hit by the pandemic within their own territories. The strong and sufficient response against Covid-19 was far from being an act of neither generosity nor any form of humanism. Other deadly diseases such as malaria[26] have been killing millions of people across the world for decades but have never received such a global mobilization to seek the treatments of these diseases; some Africans suggest that had Covid-19 not impacted these world's wealthiest nations, the mobilization against it would not have taken the dimensions it took and the vaccine would not be available 3 years after Covid-19 broke out.

That the vaccine against Covid-19 has been made in such a short period is a very good thing! With the vaccine, the world has gone back to normal living style. That has also increased the hopes that technology will help humanity sort out the climate disaster. Humans evolved to who they are today when they began mastering technology and when they began using fire[27] (energy). Since then and in varied degrees, humans have always used technology to bring every plight they faced under control. Penicillin (antibiotics) changed the life on the earth in ways that were unbelievable when it was discovered.[28] Revolutions such as the electric bulb, the Gutenberg's machine, the magnetic compass, the magnifying lenses, the steam engine, the transistors and the telegraph changed lives in multiple dimensions. Increased agricultural productivity due to advances in chemistry has enabled humanity to produce sufficient food to qualitatively feed billions of humans across the world. Marvelous scientific discoveries and technological inventions have continued to make human life better and better. Human life expectancy has increased by several decades over the 20th century when compared to what it was at the end of the 18th century. Technology has been able to prevent humans from being doomed to die from major catastrophes. Major natural catastrophes have washed away some emblematic genera and species such dinosaurs from the earth's surface, but humans persisted, even though their natural history is rather shorter than that of some other species.

The fabulous human successes in technology do, however, make it hard for people to think that solutions to the climate change crisis are not essentially and totally technological. For example, in 2014, a climate engineering conference was organized in Berlin, with the objective of globally discussing global ethical issues emerging from the adaptation to and/or mitigation to the effects of climate change. Most people who attended the conference came out with optimism that technology will be able to provide solutions humans need to solve the climate change crisis. Renewable energy to contain and gradually reduce GHG emissions is about technology. Bio-engineering to find other alternative sources of food is essentially technologically fixing climatic problems. Geo-engineering is a technological development. The first type of geo-engineering is about building a mirror-like shield above the earth so that it reflects incident solar rays back to the outer space. This would protect the earth from overheating. The second type of geo-engineering is about spraying chemicals in the atmosphere to pull GHGs present in the atmosphere down to earth before GHGs complete their residence time in the atmosphere.[29] Why could technological fixes not be possible? Humans send engines to the sky to deviate asteroids from their orbits so that asteroids do not hit the earth.[30] If humans are this ingenuous, they can, most likely, find technical solutions to climate change problems.

Without being extremely skeptical of technological-based solutions to climate change,[31] the attitude consisting of the belief that technology will solve everything is conducive to climate carelessness. All technological climate solutions may be fully understandable given the technological successes that humanity has had in its history, but it ignores the magnitude of the problem humanity faces. The belief in the technological solution prevents people from acting on their consumption behaviors and reduce individual ecological footprints by reducing GHG emissions. If there is a technological solution lurking above the heads, why suddenly change the comforts of life that humanity dearly acquired? The lure of technological quick fixes and the unquestionably brain-ingrained faith in technology slow the speed of implementing resolutions taken during the COPs. Both technological lure and naïve faith in technology are serious hurdles to achieve global climate change goals. People need some changes in their living modes.

The technological lure and blind faith in technology are not so different from the defeatist attitude where people think that climate change is so immense that nothing can be done about it. Scientific innovations and technologies are, surely, among means to save the planet from the climate catastrophe, but they are not the only ones.[32] Humanity needs to change the paradigms that have governed its living styles over centuries.

Such paradigmatic changes would combine advances in technologies and natural ecosystems that include multifunctional and sustainable biodiversity, climate and land.[32] Advances needed in technology should be climate-friendly to help solve the climate change crisis. These climate-friendly technologies can work only if they are leaned on a human ecosystem that emphasizes the need to lead change in a globally participative way. Fair industrial transformations to solve climate change will need to be global and synergic; they should be transferable to where advances in technologies are needed to support new ways of life but also to protect cultural norms as much as possible. This is what solidarity means in the era of climate change.

The discussion on Covid-19 tells a positive and a negative story in relation with climate change. The positive side is that when pushed closer to extinction, humanity can react in such a way that it tries everything to maintain itself. There is hope that despite the immensity and the novelty of climate change and social inertia that prevent humans from finding adequate solutions to the climate crisis, when pushed closer to catastrophic levels, humans would stand up together to fight the unbearable climate change. The negative side is that until the world's wealthiest nations are brought to feel a significant threat in their own lives, implementing the identified solutions to the climate change crisis will be very slow, if this happens at all. That is why developing African nations question the sincerity of the global pledges to implement decisions made during most meetings on climate change. That interrogation is rightly so when one reflects on the outcomes of most mass-gathering events of the COPs. The following chapter discusses some reasons why these good intentions are difficult to be implemented in the African context.

Notes

1 Inogwabini BI (2018) Changement climatique, un phénomène qui interpelle l'Afrique autant que le monde. Congo-Afrique 529: 808–820.
2 Meadows D, Randers J, Meadows D (2006) Limits to Growth: The 30-Year Update. Earthscan, p. 363.
3 Noga J, Wolbring G (2013) An Analysis of the United Nations Conference on Sustainable Development (Rio +20) discourse using an ability expectation lens. Sustainability 5: 3615–3639.
4 Brulle RJ, Norgaard KM (2019) Avoiding cultural trauma: Climate change and social inertia. Environmental Politics 28, 886–908. https://doi.org/10.1080/09644016.2018.1562138.
5 Collins K, Bell R (1997) Personality and aggression: The Dissipation-Rumination Scale. Personality and Individual Differences 22 (5): 751–755.
6 Chapman T (2011) Smoke and mirrors: The influence of cultural inertia on social and economic development in a polycentric urban region. Urban Studies 48 (5): 1037–1057.
7 Adams M (2013) Inaction and environmental crisis: Narrative, defence mechanisms and the social organisation of denial. Psychoanalysis, Culture, & Society 19 (1): 52–71.

8 Monroe KR, Maher KH (1995) Psychology and rational actor theory. Political Psychology 16 (1): 1–21.

9 Ball RA (1968) A poverty case: The analgesic subculture of the Southern Appalachians. American Sociological Review 33 (6): 885–895.

10 Cianconi P, Hanife B, Grillo, F, Zhang K, Janiri L (2021) Human responses and adaptation in a changing climate: A framework integrating biological, psychological, and behavioural aspects. Life 11: 895. https://doi.org/10.3390/life11090895.

11 Thomas K, Hardy RD, Lazrus H, Mendez M, Orlove B, Rivera-Collazo I, Roberts JT, Rockman M, Warner BP, Winthrop R (2019) Explaining differential vulnerability to climate change: A social science review. WIREs Climate Change 10: e565.

12 Welburn D (2018) Rawls and the Environmental Crisis. Routledge, p. 146.

13 Cianconi P, Hanife B, Grillo, F, Zhang K, Janiri L (2021) Human responses and adaptation in a changing climate: A framework integrating biological, psychological, and behavioural aspects. Life 11: 895. https://doi.org/10.3390/life11090895.

14 Grant H, Goldhamer T (2021) Our brains were not built for this much uncertainty. Harvard Business Review: Behavioral Science. https://hbr.org/2021/09/our-brains-were-not-built-for-this-much-uncertainty

15 Grupe D, Nitschke J (2013) Uncertainty and anticipation in anxiety: An integrated neurobiological and psychological perspective. Nature Review of Neurosciences 14: 488–501.

16 Collier P, Conway G, Venables T (2008) Climate change and Africa. Oxford Review of Economic Policy 24 (2): 337–353.

17 Metz B, Davidson O, Bosch P, Dave R, Meyer L (2007) Climate change 2007 - Mitigation of climate change. Intergovernmental Panel on Climate Change.

18 Nicholls RJ, Lowe JA (2004) Benefits of mitigation of climate change for coastal areas. Global Environmental Change 14: 229–244.

19 Fawzy S, Osman AI, Doran J, Rooney DW (2020) Strategies for mitigation of climate change: A review. Environmental Chemistry Letters 18: 2069–2094.

20 Sarkodie SA, Strezov V (2019) Economic, social and governance adaptation readiness for mitigation of climate change vulnerability: Evidence from 192 countries. Science of the Total Environment 656: 150–164.

21 Cucinotta D, Vanelli M (2020) WHO declares COVID-19 a pandemic. Acta Biomedical 91 (1): 157–160.

22 Guest J, Del Rio C, Sanchez T (2020) The three steps needed to end the COVID-19 pandemic: Bold public health leadership, rapid innovations, and courageous political will. JMIR Public Health Surveillance 6(2): e19043.

23 Nejati-Zarnaqi B, Sahebi A, Jahangiri K (2021) Factors affecting management of corpses of the confirmed COVID-19 patients during pandemic: A systematic review. Journal of Forensic Legal Medicine 84: 102273.

24 Or Z, Gandré C, Zaleski ID, Steffen M (2021) France's response to the Covid-19 pandemic: Between a rock and a hard place. Health Economics, Policy and Law: 1–13. doi:10.1017/S1744133121000165

25 Light DW, Lexchin J (2021) The costs of coronavirus vaccines and their pricing. Journal of the Royal Society of Medicine 114 (11): 502–504.

26 Aborode AT, David KB, Uwishema O, Nathaniel AL, Imisioluwa JO, Onigbinde SB, Farooq F (2021) Fighting COVID-19 at the expense of malaria in Africa: The consequences and policy options. American Journal of Tropical Medicine and Hygiene 104 (1): 26–29.

27 Wrangham R (2010) Catching Fire: How Cooking Made Us Human. Basic Books, p. 320.

28 Allman T (2018) How Antibiotics Changed the World: How Science Changed the World. ReferencePoint Press, p. 80.

29 This definition used examples to simplify the version of a technical and sophisticated definition found in Royal Society (2009) Geo-engineering the climate: science, governance and uncertainty. Copyedited and Typeset by Techset Composition Limited, p. 82. Despite being that simplified, however, the definition provided here captures the essential ideas of what geo-engineering, which are (1) Carbon dioxide removal (CDR) techniques which remove CO_2 from the atmosphere and (2) Solar Radiation Management (SRM) techniques that reflect a small percentage of the sun's light and heat back into space.

30 Osborne M (2022) NASA successfully crashed a spacecraft into its asteroid target. Smithsonian Magazine of September 26, 2022. Available at: https://www.smithsonianmag.com/smart-news/nasa-successfully-crashed-a-spacecraft-into-its-asteroid-target-180980840/.

31 Nisbet MC (2014) Disruptive ideas: Public intellectuals and their arguments for action on climate change. WIREs Climate Change 5: 809–823. https://doi.org/10.1002/wcc.317.

32 Ricart S, Olcina J, Rico AR (2019) Evaluating public attitudes and farmers' beliefs towards climate change adaptation: Awareness, perception, and populism at European level. Land 8 (1): 4.

2 International Good Political Intentions

Difficulties to Implement Them in Central Africa

Introduction

The proliferation of documents and treaties signed since 1992 evidences the truistic statement that solving climate change has become a fundamental quest for global human consciousness. Since that unique moment in the history of development, the world has gone from one Conference of Party (COP) to another; climate consultations are organized at both local (states and regions) and global scale; they happen in various formats. COPs are the most emblematic of these climate change mega-events regularly held since 1995. The sole operational objective of COPs remains to identify concrete solutions to combat climate change. Naturally, all COPs yearly yield lists of resolutions intended to pull the earth back from a climatic trajectory that would threaten life on the earth. COPs intend to ensure that the currently changing climate doesn't go off the climatic limits that maintain life's conditions bearable. Another way to state the sole operational objective is to qualify COPs as being part of a global process aimed at identifying sensible ways to politically and globally manage CO_2 and other greenhouse gases (GHGs). COPs are part of the international politics; they constitute high-profile diplomatic moments. If there is anything different from the conventional diplomacy, it is that COPs bring together political leaders, scientists and global civil society organizations. Climate COPs are not the only climate change grand masses that the United Nations Framework Convention on Climate Change (UNFCCC) convenes; there are also other forums that are organized in different regions and countries across the world.

Over time, the strong impressions that each COP and other forms of climate negotiation leave behind are that the resolutions they take are less practical and less applicable in the concrete daily life of humans. This is particularly for ordinary citizens of the Congo Basin. The fundamental question is why there are so many meetings, so much noise and so many resolutions for little practical progress? This chapter tackles the possible reasons for this slowness in action to execute the resolutions COPs agree on in Central Africa. The chapter does not go through all the resolutions from each COP

DOI: 10.4324/9781003493754-3

one by one; the resolutions issuing from each COP are very numerous. The chapter doesn't review the policies on climate change that each Congo Basin state has promulgated. The chapter focuses on the major difficulties to implement COP's resolutions and national climate change policies. There are four major obstacles to achieving the expected progress on climate change: (1) the notion of the right to development, (2) the principle of common but differentiated responsibilities, (3) the principle of precaution and (4) the format of climate negotiations. These sticking points are so sticky because of different interpretations they lead to. In addition, a large part of the failures to implement resolutions on climate change in the Congo Basin correlates to a lack of a science that anchors its epistemology in African realities. Because current climatic knowledge is often external to African realities, the idea of creating structures that can support research and the incorporation of the climatic question in the schemes and processes of sustainable development for Central Africa is important. However, this is not to claim that Central Africa should start building its own climate science from scratch. Ironically, the poverty levels of Central African countries offer a chance to test climate-sensitive and ecologically sustainable development because most things are to build. Before going further, briefly repeating the fact that climate change is so novel (Chapter 1) is important because this remains in the background of the discussions that follow throughout this chapter. Suffice it to be recalled in this chapter that affirming the novelty and immensity of climate change is not to think that climate change and its consequences are impossible to handle. Again, the short discussion on what should be done to deal with Covid-19 shows that humanity can break out of the traditional thinking and living ways if pushed too close to its extinction. Before tackling the three fundamental principles to which Africans cling (the right to development, common but differentiated responsibilities and the precautionary principle), some thoughts on climatic skepticism and the damage it causes in Central Africa are presented.

Half-True Climate Change Ideas Create Climatic Skepticism and Inhibit Real Action

While discussing climate change with people in Central Africa, one thing pops up consistently and is hotly debated. The thing is how the half-true ideas spread across a range of people, including the most brilliant. Climate-skeptic people create half-truths true fake news out of nothing to disparage the climate action across the world. At the other end of the range, there are evidence-based facts that are extracted from their contexts. These facts without contexts are given interpretations that are often entirely wild. Banda[1] assessed the extent to which climate change skepticism prevailed in dominant climate news in Sub-Saharan Africa and found that climate skepticism seemed limited in most African communities. Despite this good news, climate skepticism highly prevailed among town-dwelling African elites. This is not a negligible fact given the influence the African town-dwelling elites have on their rural family members and the wider flocks of rural compatriots. International news produced by international media located in the wealthiest nations of the world generates climate skepticism; this latter often feeds the African elites in major cities. Apart from the perniciousness of climate skepticism occurring among African elites, one study found that lower climate awareness was associated with authoritarian and intolerant political leaderships in Africa.[2] Populist and nationalistic ideologies nourish low climate awareness, authoritarian and intolerant African political leadership. Populist ideologies convey the message that Western countries are plotting and will always plot to maintain African countries under the yoke of colonialism (neo-colonialism) to exploit African natural resources. These ideologies often lead African populations to think that climate change is another subterfuge to keep African countries under poverty.

A more insidious half-true idea is the theory of adaptation through using traditional knowledge. Many people believe that using traditional knowledge will help African communities to adapt to climate change effects. They think traditional knowledge is one of the pillars of the fight against climate change. Certainly, the knowledge already acquired on African environmental conditions can be used to prepare communities to face climate change consequences. However, one should remember that with the novelty of the climate question and its immensity (Chapter 1 and above in this chapter), traditional knowledge alone will no longer be adequate for such a huge task. Africa undergoes changes that have never taken place before in its history. The first change is about the continent's human demography whose growth directly impacts the patterns of occupancy and usage of its geographic space. Other changes are land degradation, shrinking water levels and other biological and mineral resources. African traditional knowledge, which proved its worth during the arduous African history, is insufficient on its own to cope with the situation. Never before had Africa sheltered more than a billion people. Central Africa, for example, rarely had to face prolonged periods of dry seasons before. With such unprecedented events, the modern state's shape and the scales of the projected climate change effects impose new approaches to life and demand new solutions. Faced with such unprecedented events, clinging to traditional knowledge is almost an intellectual snobbery. The intellectual dream of paradise that Africa lost will not help. The reality is that the magnitude of the projected climate change consequences has no precedent in human history. One can say that there is really no traditional knowledge to tackle climate change. That there is no real traditional experience of coping with climate change justifies the fact that projects premised on the flawed conceptual models to collect and use the traditional knowledge to cope with climate change cannot produce outputs they expected. It is not certain that climate programs premised in these assumptions will achieve their objectives. The point is not to throw traditional knowledge away but to use it as a basis for the production and confirmation of new knowledge. It is only a new knowledge constructed using traditional knowledge that will enable Africans to cope with the harshness conditions changed by changes in climate.

The idea of using traditional knowledge to cope with climate change can also have perverse effects if communities are continuously told that all they need to deal with climate change is to revert back to their traditional practices. Communities may come to think that climate change is not that novel. If ancestors found the solutions that can help today, it must be something not as gravest as currently claimed. This thinking is prevalent across Central Africa. In many ways, it encourages people not to take drastic actions that are needed and requested of them because old solutions would suffice. Most traditional solutions are inadequate, on their own, to hold back the extreme weather patterns from hitting communities in Central Africa. For example, civil protection services that are supposed to buffer communities against the severity of extreme weather conditions consequences are nearly absent in Central Africa; relying on traditional knowledge will not help build such services. Trying to address climate change by using traditional knowledge singly is insufficient; this does not, however, demonize the epistemological foundation that knowledge is acquired through a cumulative process.[3] The issue is that the discourse insisting on traditional knowledge can retard the acquisition of a new form of knowledge required by the novelty of the situation and it inhibits the action. Communities should be told that while it is possible to use traditional knowledge, this is only part of possible solutions.

Africans and Some Provisions of the UNFCCC

The political management of CO_2 and other GHGs is legally covered by the UNFCCC. Structurally and ideally, the UNFCCC is the 1992-Agreement (Rio de Janeiro) signed by 154 States and the European Community. The Rio de Janeiro meeting aimed at bringing participants to understand the factuality, the magnitude and the potential consequences climate change has on life. A parallel objective was to think collectively of the means necessary to get the world out of the grim perspectives that the consequences of climate change augured. The first objective was fully achieved when the UNFCCC entered into force on 21 March 1994. In 2015, the convention was ratified by 195 countries. The Rio de Janeiro convention contains 26 articles, but some articles have several paragraphs. There are three articles that Africans debate in all climate negotiations. Truly and with good reasons, Africans do not want to negotiate about the three articles (principles) at all: (1) the right to development, (2) the principle of common but differentiated responsibilities and (3) the precautionary principle.

The Right to Development

A first fundamental principle for Africans is that of the right to development, which comes from the 1986 UN Declaration on the Right to Development (UNDRD). The UNDRD preceded the UNFCCC by some 5–6 years, which means that the UNDRD's effectiveness could have ceased if the UNFCCC nullified its principles. However, the right to development was fully transposed into the third principle of the UNFCCC. The 1992-Agreement principle 3 reads that 'the right to development must be carried out in such a way as to equitably meet the developmental and environmental needs of present and future generations'. In 1995, Kofi Annan declared that it was against the yardstick of the right to development that respect for all human rights would be measured. He went on to conclude that the goal of the United Nations was to achieve a situation in which all human beings are able to realize their full potential and contribute to the development of society as a whole.

Why do Africans cling so much to this right to development? It is rather futile to dwell on a question that contains its own answer. Obviously, people in Central Africa fear that climate change serves as a motive to prevent them from their own industrialization and material development. Africans legitimately fear that talks about preserving the naturalness of biomes compartmentalize them in untouchable natural habitats that are set to remain unaltered by humans. Unaltered biomes will serve as carbon storage and pumps of oxygen to the atmosphere. Africans argue that while untouched habitats play a key role in maintaining the climatic equilibrium that has supported life on the earth, they are also infested with all forms of diseases and make the majority of Africans live under unsanitary conditions. This fear Africans have doesn't seem justifiable because no one attending COPs would dare to say out loudly that the Africa must be kept underdeveloped to maintain global climatic equilibriums right. Politically correct rhetoric would never allow being this frank. However, coldly analyzing some proposals to safeguard the forests in Central Africa would justify some of these fears. There is no need to list projects and organizations by name in this context; doing so would put this chapter at the risk of being considered defamatory. However, it must simply be recognized that certain actors in Central Africa show extreme environmentalism, which can easily lend itself to the easy criticism. It has been often voiced that Western countries and the non-governmental agencies they fund use the true phenomenon of climate change for the un-avowed[4] ideological motives. For people of Central Africa, chief among these ideologies is the idea that Western countries impose on the citizens of less industrialized countries to resign themselves to the sordid conditions of nature so that climate changes as little as possible over the course of time.[5] Because of this perception, many people of Central Africa think the climate-sensitive living standards are an additional burden for the poorest of the world. This is shallow interpretation of complex climate change issues, but it gets the ground in development debates across Central Africa.

An example of such interpretation many elites loudly tell is that of the Democratic Republic of Congo (DRC). In 2022, the DRC government auctioned 30 oil and gas blocks, including in the Cuvette Centrale region. The DRC government argues it is the right of Congolese to extract these resources to invest financial income flowing from this exploitation into sustainable development activities. For biodiversity conservation international organizations, some of 30 blocs extensively overlapped with the world's largest tropical peatland, which are massive carbon sinks while another part of the blocs included UNESCO World Heritage Sites. Some of the habitats are homes to gorillas and other charismatic tropical wild animal species. For this latter group, drilling oil on Cuvette Centrale's peatlands will trigger the climate bomb loose[6] and release huge CO_2 quantities into the atmosphere. Furthermore, extracting oil in other ecosystems would jeopardize the future of high conservation concern wildlife species. The claim of the Congolese right to use the resources nature endowed their country with to make their lives better was a morally justifiable one. Unfortunately, rather than discussing better ways in which climate change and biodiversity conservation can be reconciled with sustainable development,[7] international political and economic scientists issued an alarmist outcry. The debate opposed the claim over the right to development[8] against the universal value of preserving CO_2 sinks intact and wildlife species habitats functional. Realistically, this episode was simply misplaced because the question should not have been if either some people have right to development or they were doomed to remain poor to preserve the world from climate cataclysms. The right question should have been how to use natural resources in the DRC to ignite a truly sustainable development for its people and without relying on CO_2-emitting technologies. Most sustainable development actors in Central Africa viewed the debate as a testimony to how Western countries impose poverty on Africa. They argued Western countries impose poverty in Central Africa to keep the Western lifestyle thriving. Even the most nuanced Congolese intelligentsia negatively viewed the intervention of biodiversity conservation organizations on this issue. Congolese who opposed the neo-colonialism narrative were a thin minority who, mostly, worked with international biodiversity conservation organizations. If the question was framed such that it evolved around how to reconcile both demands, the majority of people would have a totally different opinion. Generally, this case leads to lesson that because people are often excluded from the climate change and biodiversity conservation debates in Central Africa, authoritarian and populist ideologies acquire deep roots. These ideologies become more audible since scientific data and discourses it produces are essentially Western driven. Even African scientists are Westernized African elites.[9]

However, there is also hope in this situation because being audible is not necessarily being plausible. African communities are less naive than often imagined; they don't easily move by the first siren song. Even in this oil-climate change saga, Congolese denounced the hypocrisy of the Global North that asked Congolese to stop extracting natural resources to resorb Western climate historical responsibilities. Despite the pledges made at the COP26,[10] Congolese know that the same Western countries continue emitting GHGs; some countries, such as the United States, continue producing fossil fuels through fracking. Congolese also questioned why these zones were auctioned; many suggested that auctioning large expanses of the national territory was for politicians to amass money one year before elections. Because of the long history theft of the country's assets by its political leaders, the majority of Congolese argued that the oil auction gamble was not caused by a genuine desire to develop the country; they felt the gamble was to prepare political retreat for politicians who would lose the elections. Otherwise, they thought the money from the oil auction would provide significant resources to the incumbent political leadership to finance its electoral campaign. Notwithstanding the Congolese debate, Central African countries sided along with oil-producing countries at the COP28 to smartly avoid endorsing the idea to phase out from fossil fuel, which shows their faith to stick to the principle of their right to development.

Principle of Common but Differentiated Responsibilities

The principle of common but differentiated responsibilities is the 1992-Agreement principle 7. It is based on the idea that it is unfair to subject the less industrialized countries to the same environmental obligations as the more industrialized countries. Industrialized countries should accept their historical and current responsibilities in the international effort to support sustainable development. Morally, principle 7 grounds on accountability for the pressures industrialized societies exerted and continue to exert on the global environment since industrialization began. It is about historical responsibilities on which less industrialized African countries insist to obtain financial contributions from industrialized countries. Principle 7 by its moral nature is not a simple demand for solidarity. It is not because most of the financial assets are readily available to industrialized countries that African countries demand for solidarity, as extreme right politicians often present the issue.[11] Principle 7 is about climate reparations; it is a principle of justice. Justice is a claim that everyone is held responsible for the actions that are attributable to each. Of course, the attribution of climate change causes raises hard questions that the COPs find difficult to sort out.[12] While Africans call on industrialized countries to bear a larger financial share of the burden to correct global climate actions, some Westerners think current Western generations are not co-responsible for the past action of their ancestors. Younger generations are not duty-bearers for past actions their ancestors have taken; they should not pay for the wrongs they did not cause.[13] This is not the co-responsibility rebuttal argument with which it is unsurprising that financial promises made during the COPs take too long to materialize. The 'no co-responsibility rebuttal' has one big logical flaw. The current generations in industrialized countries inherited the wealth that was created under historically unjust[14] conditions. They cannot take and continue to use the assets generated under these conditions while fighting to get rid of the liabilities the production of these assets led to.[15] Assets generated during the past climate unjust conditions have continued to produce benefits that the younger generations to take advantage of. It is morally plausible to ask those who currently own these assets to contribute to the price of fixing the climate and the price to allow the climatically most vulnerable communities to adapt to the new normalcy.

Pricing the Action against Climate Change

The proponents of the 'no co-responsibility rebuttal' sometimes make raucous noise against historical responsibility. However, in 2009,[16] most industrialized countries agreed to contribute 100 billion dollars annually to finance the climate action in the least industrialized countries through 2020. It is currently known that this amount has never been disbursed before 2020. The promise was renewed in 2015 (COP 21, Paris) to disburse the same amount annually until 2025. The figures currently made available to less industrialized countries under this pledge remain difficult to tabulate. This difficulty comes from the fact that certain amounts of bilateral and multilateral aid have been rebranded. Some debt remissions were also transformed into climate action package (Figure 2.1[17]). Despite the hopes the promised 100 billion dollars raised the financial effort to support the climate action is below the expected

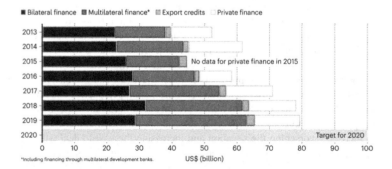

Figure 2.1 Types of aid funding allocated to less industrialized countries to combat climate change and the gaps for achieving the USD 100 billion per year objective between 2013 and 2020.

Source: Jocelyn Timperley, cited above.

amplitudes. Genuinely, the political will to move forward with the climate action is weak. Three realities would explain this state of affairs.

First, when reading Figure 2.1, financial mechanisms to fight climate change draw are not so different from traditional international financing mechanisms. Using traditional bilateral and multilateral aids, there is no fresh capital but only a realignment of traditional aid funds to the new realities. Second, the USD 100 billion had to go to 54 African countries and other less industrialized countries. Simple arithmetic shows that even if the division was fair and was only between African countries, each would have about USD 1.9 billion annually. USD 1.9 billion annually would certainly provide an economic boost for many African countries. But Africans often ignore the fact that the theoretical amount is not intended only for African countries. The USD 100 billion are also for other under-industrialized countries. There are many less industrialized countries that expect funds from the same envelope to support their climate change actions. African countries should lower the hopes they placed in these 100 billion US dollars. According to Voïta, only 3% of the USD 100 billion flowed annually to Sub-Saharan Africa[18] in 2023. What is more, the USD 100 billion is taken as help. As aid, there would be no constraints for donors to fulfill their promises because aid is solely at their discretion. This helps to understand that barely a meager portion of this funding was delivered before 2020. Third, realism dictates a cold reading of the magnitude of the USD 100 billion and put it into perspectives related to the extent of the task that awaits less industrialized nations if they are to work effectively against climate change. Without much difficulty, it is observed that this amount could not replace the costs of basic needs of African populations for an effective fight against climate change. Combating climate change requires profound changes in the lives of Africans. Such weighty changes in Africans lives will have a high cost. Of course, everything should not be paid for by Western industrial countries. Even if those who benefited from unjust past acts and injustice should pay for climate, it is pushing the argument to absurdity to ask Western countries to pay the price alone. The costs for climate action are very high; Voïta has even projected that the current levels of funding would need to increase by at least 590% annually to address the current needs.

According to Voïta, the prospects to meet the USD 100 billion promise looked encouraging by the end 2023. This promising perspective should not hide the fact that the USD 100 billion promise was made since the COP 15 (Copenhagen) more than two decades ago and that GHG emissions continued to increase, worsening the situation and increasing the cost to deal with the climate crisis.[19] There is a need to envisage other funding mechanisms. These mechanisms should not rely solely on international aid. One such other mechanism is the Loss and Damage Mechanism (LDM), which should help people repair damages they incur as consequences of climate change. The LDM was established at the COP 19 (2013, Warsaw).

The LDM was among the sticking points of COP 27 (Sharm el-Sheikh) with no major breakthrough. It was a surprise for African countries and the Global South that COP 28 (2023, Dubai) began with the announcement that the LDM funds have been established. Unfortunately, the happiness the announcement raised was short-lived because the amount pledged was by no means near what has been expected by those who incur climate losses and damages. Beyond the fact that the amount committed was deceivably meager, the process to access the funding allocated to LDM funds raised concerns, as was the case for its hosting institution. Ironically, the LDM is a voluntary mechanism whereby wealthy nations will contribute according to their willingness. The justification for this awkward situation is framed following the attribution question; methods one country could use to unquestionably attribute the losses and damages its people incur to any particular developed country are difficult to ascertain. All this shows that African countries need to bear their equitable share of the burden as well; how they can arrive at this is discussed in Chapters 7 and 8.

To increase the pledged 100 billion USD fund annually by 590% to address the current needs confirms the claim that the 100 billion USD was not sufficient for African countries to cope with climate change. To covert the energy bill to transition one third of Congolese from charcoal to clean energy, for example, has been estimated to cost USD 10.5 billion[20] of initial investment, which means that one would need roughly around USD 31.5 billion to cover the entire country. Given how difficult it is to access the USD 100 billion climate fund, the DRC will not move to cleaner energy soon. Beyond the funding question, to get the DRC out of being totally charcoal dependent to cooking would mean not only improving the supply of clean electricity but also compensating for jobs for charcoal sellers and other services that contribute to the economy of the charcoal market, including charcoal packaging, its transport, charcoal handling and charcoal storage. These services were not included in the calculations that led to the figure above. USD 31.5 billion investment is about a third of the entire climate 100 billion USD fund. The use of charcoal is not the only source of GHGs in the DRC. Transport, industrial facilities and obsolete machineries, for example, need to be transformed to become CO_2 (GHG) neutral or at least less consumptive. A complete renewal of the industry and replacing roads by rails, like the example of shifting from charcoal to cleaner energy sources, require more money and will certainly take time to materialize.

Should Climate Payments Be Considered International Aid or Justified Obligation?

The demand developing countries formulate to industrialized countries to pay to change their living ways constitutes a Gordian knot of all climate COPs. Certain citizens of developed countries feel less co-responsible for the action of their ancestors, and they don't think they should be channeling large sums of funds to less developed countries to curb or adapt to climate consequences. The lukewarm appetite to transfer funds to poor countries under the climate payment mechanisms can be explained by the fact that most Western countries consider that poor countries claim amounts that are already part of the normal bilateral and multilateral aid. The aid is neither a formal moral nor a legal requirement; aid is not a right, and it should go where the donors want[21] it to go and should cover the areas the givers desired to cover at the pace they want it to be disbursed. The climate change fund falls within the logic of international aid. Financial aid is to continue supporting the development of donor countries; it follows a pattern where there is a positive loop whereby donor countries export goods to recipient countries and recipient countries import the same goods using climate aid transfers.[22] The positive correlation between climate transfers and the exports from climate donors and their imports from climate fund recipient countries shows that what determines the flow of international climate finance is neither the capability to pay nor the ethical compulsion to do so but rather self-interest of the donor countries.[23] People from Central Africa read this situation with the lenses of postcolonial theory[24] when they see that the donor's self-interest dictates the pace, the amount and where the climate funds go. Aid, even with no obvious donor's self-interest, comes as paternalist help[25] in which donors pretend to know the communities' needs better and more than communities themselves. Sadly, communities in Central Africa conclude that aid, the global aid, is to help the donor countries to maintain the former colonies under the yoke of their former masters. Aid also weakens African states in that they make little effort to take care of the needs of their fellow citizens.[26]

The demand for climate funds made by Africans and other poor nations is not as far-fetched as it may seem. African countries and other poor countries are entitled to demand funds because their demand is a case of justice. The principle to hold everyone responsible for their actions should apply in this case. The 'no co-responsibility rebuttal' hides lack of political willingness and justice as fairness. The 'no co-responsibility rebuttal' can only hold where people want to keep their historical privileges. Proponents of the 'no co-responsibility rebuttal', after benefiting from past unjust acts, are willing to continue benefiting from injustice.[27] The destructive action of climatic conditions did not stop until the beginning of the industrial era; it has continued to the present day in developed countries. Seeking reparation is not tantamount to appealing to a debt of the past; it is to appeal to a fair redistribution of the weight of the burden of repairing the climate weighted by the contribution to the causes that led the world there. Even if it is sometimes made in somewhat folkloric ways, the request for developed countries to pay for climate is also morally right. The correctness of this demand finds its relevance in justice which, as Aristotle said, was 'a matter of proportion'.[28]

The current international climate financial assets are part of the ambivalent international relations. International relations are necessarily made up of imbalances in political, economic and military power. It is not aberrant that Africans are standing up and demanding collective action to work on the climate change causes. People from Central Africa don't ask for alms; they ask for just reparation: this is to stand for a just and noble cause. Even the majority of people in the developed countries are cognizant of this fact and would pledge to come forth with support, at least verbally. The surprise, for most people in developing countries, is that promises are not kept. The reason why the old thinking and mechanisms of international aid are kept going is, again, social inertia (Chapter 1), which can be equated to political inertia at this point. Political inertia is the resistance to changes in politics; shifting acquired political attitudes is not that easy.

Precautionary Principle: When Understanding One Concept Shifts by 180°

When discussing climate change in Central Africa, many interpretations are given to principle 17 of the 1992-Agreement. Principle 17 says that in the event of risk and in the face of the deterioration of the environmental situation, the lack of certainty in scientific knowledge does not justify inaction. This principle is interpreted in several ways in Central Africa. These interpretations range from the demand for total inaction to the demand for excessive action. The proponents of total inaction argue, for example, that it is necessary to wait until total scientific knowledge is available to begin a lucid exploitation of forest ecosystems. An example where debating different interpretations of this principle emerged in the discussions about the 30 blocks of land that the DRC auctioned for oil discussed above. Activist biodiversity conservation and climate organizations used the precautionary principle to ask the officials to wait for knowledge before any action was taken. For the DRC Minister of Environment and Sustainable Development, it was not because everything was not known that the DRC could not use its natural resources to develop. In some sense, they were arguing the other way around; claiming that the lack of knowledge may plead for their action. Those who demand excessive action believe in action, even when there is no evidence to support why there is a need for the action. For example, there is a group of climate activists who think that foresting all areas of the continent is needed to act on climate change. They suggest this to be the case even in ecosystems that were not 'naturally' forested over long-term ecological history. This demand only translates the interpretative limits of the precautionary principle. Ecologically, it makes no sense to suppress the diversity of habitats, as part of what biodiversity is, to solve the climate crisis. What is needed is to keep biodiversity functional and climate under bearable conditions. These two cases are consequences of limited academically qualified capacities in

Central Africa working on a coherent climate program. Most people add pell-mell everything together without in-depth knowledge. Unquestionably, there are some individuals qualified in climate sciences, but they are not enough.

The Format of the Climate Negotiations:
Diplomacy as Usual

Climate change negotiations engage several parties, including political, state, private and international civil society actors. Climate change negotiations put citizen participation's principles in practice; everybody should rejoice on this. To exult is an appropriate sentiment because climate change is a very sensitive issue; it would be logically and politically bad if a minority of the world's internationalist political and financial elites took climate's decisions alone. Likewise, everybody can be happy to see climate negotiations are as inclusive as possible. However, it remains true that each stakeholder who goes to the time COPs has their own agenda and methods. This is normal; each party can bring only its claims and red lines that it must confront with those of the others. But climate negotiations take place between diplomats with well-honed languages full of stylistic hyperboles, scientists whose languages is often conditional but sometimes without gloves, activists for whom the ideal would be summed up in their slogans or in the sound of vuvuzelas, governments-sponsored tourists and tourists supported by other international organizations and diverse groups of different social movements. This diversity of actors often makes climate negotiations something other than negotiation: it is cacophony.[29] Often COPs create an environment where everyone talks but no one follows, except when delegation with economic and political might speak. This environment conduces to negotiations that ultimately make participants leave disappointed with the results obtained during the discussions. It can only be disappointing because it is often behind the scenes of the amphitheaters[30] that the diplomats of the industrialized countries finalize the decisions the plenaries adopt. Plenary rooms are filled with participants who commonly share only the fact of being together in the same meeting room. Despite the remarkable impression of total open citizen participation, finally, climate decisions are mostly decisions taken in totally classic diplomatic negotiations. Climate diplomacy is not a renewed form of diplomacy; it is still diplomacy. Unsurprisingly, climate negotiations' decisions are not that extraordinary; the only things that are unusual are the lyricism that characterizes the naïve participants' language and the applause that follows the adoption of decisions they reach.

This chapter discussed why it feels like climate change COPs have intangible effects. I argued that major obstacles that lead to lack of progress are in the conception of the notion of fighting climate change effects. Mainly, the right to development, the principles of common but differentiated responsibilities, the precautionary principle and climate negotiations' format constitute obstacles to progress. Rather than being a new diplomacy's form, climate diplomacy remains the same old-style international politics. In such conditions, how are the power imbalances to be addressed given that climate change places all the nations at the same lot? This point is taken up for the specific case of the DRC, used to demonstrate what can be sensibly done for Central Africa.

Notes

1 Banda SC (2021) Assessing and evaluating the dominant themes of online climate change news coverage in Sub-Saharan Africa to ascertain the prevalence of the climate change skepticism. A report of the Cornell Alliance for Science, Cornell University.
2 González JB, Sánchez A (2022) Multilevel predictors of climate change beliefs in Africa. PLoS ONE 17(4): e0266387. https://doi.org/10.1371/journal.pone.0266387.
3 Lee J (2012) Cumulative Learning. In Seel NM (Editor) Encyclopedia of the Sciences of Learning. Springer.
4 Ferry L (1992) Nouvel ordre écologique: l'arbre, l'animal et l'homme. Editions Grasset & Fasquelle, p. 222.
5 Harvey D (2003) The New Imperialism. Oxford University Press, p. 263.
6 Carrington D, Taylor M (2022) Revealed: The 'carbon bombs' set to trigger catastrophic climate breakdown. The Guardian, 11 May 2022.
7 Inogwabini BI (2007) Can biodiversity conservation be reconciled with development? Oryx 41 (2): 2–3.
8 Mongabay (2022) In the DRC's forests, a tug-of-war between oil and aid. MONGABAY of 09 June 2022.
9 Grove R (1995) Green Imperialism: Colonial Expansion, Tropical Island Edens and the Origins of Environmentalism (1600-1860). Cambridge University Press, p. 560.
10 Civillini M (2023) Rich countries sink billions into oil and gas despite COP26 pledge. Climate Home News, 07 September 2023.
11 Jagers SC, Harring N, Matti S (2018) Environmental management from left to right–on ideology, policy-specific beliefs and pro-environmental policy support. Journal of Environmental Planning *and* Management 61 (1), 86–104.
12 Van Oldenborgh GJ, Van der Wiel K, Kew S, Philip S, Otto F, Vautard R, King A, Lott F, Arrighi J, Singh R and Van Aalst M (2020) Pathways and pitfalls in extreme event attribution. Climatic Change (2021) 166: 13.
13 Barry C, Wiens D (2016) Benefiting from wrongdoing and sustaining wrongful harm. Journal of Moral Philosophy 13: 530–552.
14 Berkey B (2017) Benefiting from unjust acts and benefiting from injustice: Historical emissions and the beneficiary pays principle. In Meyer LH and Sanklecha P (Editors) Climate Justice and Historical Emissions, Cambridge University Press, pp. 123–140.
15 Baatz C (2013) Responsibility for the past? Some thoughts on compensating those vulnerable to climate change in developing countries. Ethics, Policy, & Environment 16: 94–110.
16 UN Climate Action, Finances & Justice. https://www.un.org/en/climatechange/raising-ambition/climate-finance
17 Timperley J (2021) The broken $100-billion promise of climate finance and how to fix it: At Glasgow's COP26 summit, countries will argue for more money to mitigate and adapt to the effects of climate change. Nature 598: 400–402.
18 Voïta T (2023) Is international climate finance unfair and inefficient? Briefings de l'IFRI, p. 16.

19 Henry C (2023) Pour éviter un crime écologique de masse. Odile Jacob, p. 239.

20 World Bank (2020) Increasing Access to Electricity in the Democratic Republic of Congo: Opportunities and Challenges. World Bank, p. 88.

21 Moyo D (2010) Dead aid: Why Aid Is Not Working and How There Is a Better. Farrar, Straus and Giroux, p. 208.

22 Bayramoglu B, Jacques JF, Nedoncelle C, Neumann-Noel Lucille (2023) International climate aid and trade. Journal of Environmental Economics and Management 117 (102748). https://doi.org/10.1016/j.jeem.2022.102748.

23 Qian H, Qi J, Gao X (2023) What determines international climate finance? Payment capability, self-interests and political commitment? Global Public Policy and Governance 3: 41–59.

24 Abubakar D (2015) Africa in world politics and the political economy of postcoloniality. In Omeje K (Editor) The Crises of Postcoloniality in Africa, The Council for the Development of Social Science Research in Africa (CODESRIA), pp. 45–64.

25 Tomasello M (2016) A Natural History of Human Morality. Harvard University Press, p. 194.

26 Calderisi R (2014) The Trouble with Africa: Why Foreign Aid Isn't Working, New York City, St. Martin's Press, p. 256.

27 Berkey B (2017) Benefiting from unjust acts and benefiting from injustice: Historical emissions and the beneficiary pays principle. In Meyer LH and Sanklecha P (Editors) Climate Justice and Historical Emissions, Cambridge University Press, pp. 123–140.

28 Aristotle (2004) The Nicomachean Ethics (Translated by Thomson HT). Penguin, p. 400.

29 Huet S (2015) Les dessous de la cacophonie climatique. La ville brûle, p. 144.

30 Drexhage J, Murphy D, Brown O, Cosbey A, Dickey P, Parry JE, Van Ham J, Tarasofsky R, Darkin B (2007) Les changements climatiques et la politique étrangère Exploration des options de meilleure intégration. Institut International du Développement Durable, p. 88.

3 How Can Central Africa Successfully Play Its Global Political Role in Managing CO_2 and Other GHGs?

Introduction

The end of Chapter 2 suggests that most of developing nations play a sheep-follows-the-sheepherder role in climate change discussions. Most developing countries have very little influence on the course of things because they follow world leaders in economy, politics and science. Some climate change negotiations' inherent problems described above are a common knowledge. They are world's usual political power struggles whereby the politically weak are forced to become meek and adopt whatever the most powerful ones impose. The ordinary power imbalances that transpire from Conference of Parties (COPs), with diplomacy remaining usual, take the frame of constrained political mechanisms.[1] Politically constrained mechanisms are made of principals (bosses) and agents (subordinates). Generally, bosses expect subordinates to do good jobs. Industrially advanced countries act like the bosses expecting the subordinates (developing nations) to act in such a way that needs to address vexing climate change questions are sorted out without touching on the current political and economic equilibriums. These questions can be addressed and fixed if there is good will in parties negotiating on climate change. Unfortunately, despite the danger climate change represents, no one can expect abrupt metaphysical changes in ways in which humans conduct their affairs (Chapter 1). Diplomacy remains and will continue for several decades to be dictated by national interests. For climate change negotiations to achieve meaningful and balanced outcomes, some pre-requisites are to be put in place in developing nations. These pre-requisites will allow developing countries to form coalitions. Political coalitions are a model suggested to transition political power imbalances to a global world democracy. The world needs new power structures and global democracy because climate change is a global crisis. This chapter presents and discusses these pre-requisites for Central Africa. In the following paragraphs, Congolese means citizens of the DRC, and the DRC is cited as an example and should mean Congo Basin by extrapolation unless otherwise stated.

DOI: 10.4324/9781003493754-4

To discuss of some of these pre-requisites, it is useful to come back to the main idea of Chapter 2. Climate change eruption in the world management by humans has significantly modified the relationship between humans and nature. Climate change will characterize of all life's explainable paradigms (cultural, economic, political and social) in the 21st century. This means there will be talks about culture, civilization, economy and development without thinking about the thorny climate change problem. This is why fighting climate change harmful effects is a prime political issue because climate change threatens the human civilization. But one can deal with climate change only in a global framework. The United Nations Framework Convention on Climate Change (UNFCCC) currently stands for this global framework with the overarching aim to lead the world's collective work. This much is self-evident on its own; however, efforts that will positively influence global climate change are those that take place at the local and national levels. Adding local and national actions will make the global action. Each country must give the best of its citizens' work to contribute to the mitigation of and the adaptation to climate change.

Roles and Moral Responsibilities of the Congo Basin's Countries in the Climate Action

Each Central African country must go through the UNFCCC concept and define its own practical schemes to make its contribution to the whole world climate change action. This has to be done while each Central African country also strives to ensure its people are sustainably developing. This chapter uses the DRC as an example to elaborate on how Central African countries would structure themselves to play a key role to stabilize climate degradation. Congo Basin's countries still have relatively functional ecosystems; the world expects them to become an active player in the fight against climate change.[2] What this chapter says on the DRC should apply, with some contextual adjustments to many of the African and Central African countries.

To state that the DRC should play an active role and make its contribution in efforts to address climate change recalls three things. First, the place its geography occupies in Africa positions the DRC to bear moral responsibilities for managing its environment soundly to ensure that global environmental balances are kept safe. This certainly raises questions about what type of moral responsibilities one talks about and what grounds these moral responsibilities have. An elaborated argument on these two questions falls beyond the scope of this book; producing such an elaborated argument would need an entire book of its own. However, the DRC bears moral responsibilities because its geography positions it in a place where to care about its environment for the sake of greater global good makes sense on the argument of sharing common goods.[3] Sharing goods of greater importance when one obtained these goods for free is a sensible thing to do to enable common life to continue. There are economic[4] and moral[5] arguments against the morality of 'freely ye have received; freely thou shall give'. But what maintains life on earth is beyond economic arguments because the stakes of climate change are so high; they threaten the life (Chapter 1). No sensible person would be willing to lose their own life and that of their descendants for lack of action. Clearly, the present climate situation espouses Peter Singer's drowning child argument,[6] with the truth that the drowning child in question is everyone; facing climate change consequences, the drowning child is the entire humanity, if not life itself entirely. Sitting inactive where solutions to climate change can be achieved is clearly immoral because the DRC could play an important partition that can contribute to saving the world from drowning into the climate crisis. If the DRC chooses not to act, as Peter Singer argued, it would simply be immoral. [Democratic] societies cannot be indifferent to the common life's character.[7] Ecological functions that DRC's ecosystems (forests, savannahs, lakes and rivers) play need to continue performing to ensure that services rendered to humanity are maintained sufficiently safe. Whether the DRC is democratic society or not is not the issue at the moment; it is discussed in Chapter 8.

Second, despite being naturally endowed with significant potentials to contribute to the climate change solution, the DRC is also one of the climatically most fragile countries; it is even probably the most climatically vulnerable country.[8] In Central Africa, the DRC is followed by Burundi and the Central African Republic. To a large extent, the countries of the Congo Basin are largely vulnerable to the effects of climate change. It is somewhat sad to hear only the story about the DRC being capable to provide climate change solutions. The vulnerability of ecosystems and Congolese people is often neglected. Being climate resilient is not singly a matter of ecosystems where people happen to reside: climate resilience depends on human capacities to deal with adverse climate change effects and on the institutional makeup of any given country. The political systems operating in each country are so important to ensure that human communities as well as ecosystems are well taken care of (Chapter 7).

Third, the DRC must honor commitments it made many times before the international scene. However, the DRC must not only ensure that its international commitments are kept but it must also ensure that the fulfillment of its international commitments on climate change goes hand in hand with its responsibilities toward its citizens to integrally develop the country. With climate change threatening humanity, integral development can only make sense if it is based on the matrices of sustainability. Sustainability matrices require that economic development be socially just and equitable and sensitive to ecological and environmental balances. It is to help the DRC to harmonize its vision of sustainable development and honor its international commitments on climate change issues that the chapter proposes that the DRC be able to set up a national institution whose motivations and objectives would look like what is discussed below.

Motivations to Create CO_2 and Other GHGs Political Management Structure

The DRC, as is the case for other Central African nations, does not have a national technical body in charge of climate change issues; instead, there is a cohort of public institutions scattered across ministries and private associations. Some of these bodies work on separate bits necessary to help meet the DRC its international demands, and Congolese climate demands, to a lesser extent. These bodies are accountable to the ministry in charge of the environment; they help the ministry to prepare for international sensational events and write and submit reports. These scattered efforts don't have an internal strategic coherence that can lead the DRC to fully play its role. Consequently, the DRC is unable to pull the maximum benefit from its climatically strategic position. Thus, the DRC government does not have control over the initiative to combat climate change. A careful analysis of how things go gives the impression that the DRC government is increasingly driven by national and international non-profit associations instead of being a driving force for its climate partners and allies. This doesn't imply that the role of these associations is bad in itself. However, non-profit associations pull in several, sometimes opposite, directions. Non-profit associations most often advocate for interests that can be internationally funded. With this, the DRC's partners often take different (often) opposite directions, which scatters efforts and makes the impact difficult to measure. The real impact of climate investment is challenging to ascertain for ongoing projects in the DRC. Additionally, as each initiative pulls in its own direction, the fact that the DRC does not have its own national technical climate change body means that the government has little adequate expertise equipped with scientific knowledge to inform political public action climate policies. On climate change issues, the DRC is still sailing by sight at best. In this context, the DRC government should regain the initiative and become the driving force behind climate change action. To achieve this, a national technical climate change body is essential. But this technical body should not be a politically driven corps but rather truly technical, with competencies from different horizons. Ministers should be left to organize climate change politics using the expert knowledge they get from such a technical body.

The Objectives for a Structure to Technically Manage CO_2 and Other GHGs

The strategic and specific objectives of this technical body stem from the above diagnosis. Three strategic objectives each associated with its specific objectives would be assigned to three components of a sound structure. The first strategic objective is to serve as a technical body of the DRC government for its climate change action. This strategic objective is translatable into the following specific objectives. First, to provide the most advanced technical information to the public policy-makers to help them formulate national climate change policies. Second, to prepare technical notes to clarify and support the opinion of Congolese public action vis-à-vis state and non-state partners during international climate change discussions. What Central Africa needs is what Gabon has been done. In Gabon, most efforts to address climate change were, until recently, scientifically evidenced. This situation was so because former Gabonese Minister of Water, Forests, the Sea, and Environment is an experienced researcher with a strong field experience having conducted research in Cameroon, Congo, DDRC, Uganda, Sierra Leone and Nigeria. Professor Lee White was so habile as to use his scientific strengths to slip scientific knowledge to support climate change ideas and actions the former Gabonese President would take. The Gabonese position in climate change negotiation was often the most logically constructed in Central Africa because it was from an evidence-based system. This does not suggest that everything done in and by Gabon is scientifically sound and that other countries of Central Africa should admire it. Far too is to suggest that all the Central African ministers of environments should be academics. Everything being kept equal across different contexts, the idea is that scientific evidence should guide climate change discussions in the DRC. Second, such a technical structure should assist the government to prepare, set up and implement a national climate framework. This framework is to have clear objectives and methodologies to collect data to evaluate the actions carried out in the DRC by various actors to combat climate change. Collecting scientific evidence should be part of this as well.

The second strategic objective is to harmonize all actions by all actors involved in managing the effects of climate change and preventing further deterioration of the world climate. Specific objectives flowing from this strategic objective are the following. Primarily, to inventory all local, national and international actors working on climate change, what types of actions are necessary to ensure that their respective actions contribute to the achievement of the nationally set objectives to mitigate climate change and help people adapt or become resilient to the effects of climate change. Second, the national climate change technical body is to bring the much-needed evolving and novel technical and scientific knowledge to local and national climate change actors. Sharing technical and scientific knowledge with local and national actors is necessary so that the climate change question is owned by all. The owning climate change should not be limited to the state administrations; it is not even a particular question of the Ministry of Environment. All national institutions in their diversity should embrace this question. It is through this broad ownership of climate change issues that the national policies to combat climate change will become easily understood, accepted and implemented. How do you ensure that all DRC's communities own the climate question? The short answer is that this will come only through education and popular awareness sustained over significantly long time. Given that currently there is a lack of coherence in the national climate change policies, efforts should be invested from scratch.

The third strategic objective is to produce endogenous climate change knowledge. This knowledge should fine-tune the formulation of national climate policies. DRC national climate change policies are to be rooted in national realities and national varied ecological contexts. Traditional knowledge will not suffice to curb down the effects of climate change; it needs to be re-evaluated, and its sound portions should be incorporated in the bulk of the global knowledge, where appropriate in this mega effort to deal with climate change. Traditional knowledge is not a climate change panacea. Producing current endogenous knowledge should not be equated solely to looking backward in the past. It must also be acknowledged that most concepts used to discuss climate change are absent in the memories of local communities. As an example to support this statement, between 2005 and 2010, I managed a program in the Lake Tumba Landscape (80,000 km², with 2,500,000 people) in the DRC (at the border with the Republic of Congo[9]); I commissioned a socioeconomic assessment to be carried out throughout the landscape to ensure that needs of communities were factored in plans to manage natural resources. The assessment covered about 30% of this population[10]; surprisingly, climate change themes were not evident in the languages of the region. Extreme weather events such as prolonged droughts and long massive flooding periods were recorded, but long-term trends were not. This does not mean that the people have never heard of climate change; even local researchers recruited for this project relied on detoured metaphors to speak of climate change. Extensive reliance on detoured metaphors to speak about climate change was common in this region.[11] Poor vocabulary of languages in the region doesn't cause the lack of climate change-specific vocabulary in such a large area of Central Africa, as would be thought.[12] This example supports the argument that the novelty of the climatic situation (Chapter 1) makes it difficult for people to think of it in common languages. In the 1990s, I collected different names of insects in Lingombe, a language spoken in the north-western DRC. I shared this data with Delphi Messinger[13] who immediately noticed that the majority of species that people listed were the ones that bite and cause harm to humans. This example lends support to the argument that when people have known phenomena in their collective history, and particularly things that have impacted their common well-being, they tend to be prolix in words, idioms and images of such life-impacting occurrences. This equally supports the climate novelty argument. Let recall that one objective of the climate change structure is to produce current endogenous knowledge. The work to do would also be to participate in the creation of climate specific vocabulary for contexts where there is nearly none. There is a reason to invalidate the call for forming appropriate local climate-specific vocabulary. Languages are live objects that are continually constructed and enlarged when novel conditions bring new matters to think about and new objects to be used. The information technologies provide an example where languages in Central Africa have had to create new vocabularies. The search for endogenous knowledge is also, and primarily, about identifying current practices that local communities use across the DRC to handle extreme weather events in their daily current lives. Such common extreme weather events are droughts, abnormally heavy rainfalls and atypically heat waves. The following specific objectives ensue from this third strategic objective. The first specific objective is to carry out cutting-edge climate change studies and share these studies with Congolese. Because science is essentially universal, the cutting-edge knowledge produced in the DRC should be shared with the world.

Striving to produce a localized yet universal knowledge, the technical climate structure should assist the DRC government to formulate an academic climate change program. This academic climate change program should be integrated into the curricula of secondary and university education. This suggestion is based on the reality that limited climate science capacities commonly characterize Central Africa, and this limitation in capacities is widely acknowledged. I collected data for the Sustainable Development Solutions Network (SDSN[14]) to produce a first preliminary comprehensive list of scientists working in the Congo Basin to help launch the Science Panel for the Congo Basin (SPCB) Initiative; that data supports the fact that the pool of climate scientists is very limited. Indeed, the assignment combed several methods to collect the names of scientists conducting research in Central Africa. Three-question questionnaires were sent out to 60 researchers who have been doing research in the Congo Basin over the last two decades. I asked the recipients to share the names of people everyone knew conducted research in the Congo Basin region, with expertise and email addresses where possible. Responses to this question generated a 4000-name long list. I had to sort out who to include and who to leave out in the list of researchers. I purposely decided to sieve the long list through putting a threshold of including only people whose names appear on peer-reviewed papers after 2015. To ascertain the areas of expertise of each researcher, two methods were used. First, areas of expertise comprised the emails I received back. Second, profiles of researchers were searched in public domains, including department or faculty's websites and auto-profiling webpages such as Academia, LinkedIn and ResearchGate. Many researchers have many areas of expertise; I decided to limit my analysis only to the first three areas. From the 4000-long list of names, sieving the list using peer-review publications, only *ca.* 400+ names made it through to being considered part of the scientific research community in Central Africa.

A more general excursus of this assignment is important because it gives a broad overview of the regional scientific research in the Congo Basin. The first finding was that research in the Congo Basin is mainly biodiversity conservation research, followed by soil chemistry and evolutionary biology.

Few publications emerged on other important themes such as agriculture, epidemiology, remote sensing, plant systematics, and human demography. Lacking studies on human demography was a surprise in a region where the human populations grow rapidly. That the majority of researchers were field biodiversity conservationists was not surprising because the research in question is primarily initiated by international biodiversity conservation non-governmental organizations. Non-governmental organizations fund most of the research in the Congo Basin. Researchers based outside of Central Africa (the United States, the UK, France and Belgium) dominantly conducted climate research (atmospheric sciences, core climate physics and atmospheric chemistry). Interestingly, most of the Central African atmospheric scientists were found in Cameroon, but they represented 0.1% of the 400+ peer-reviewed published scientists. The message of this excursus beyond the central argument of this book is to show why it is critically important that the proposed structure to support the government be created to draw and help implement climate change academic programs to fill the existing gap.

A paramount objective of such climate technical body is to contribute to the creation and dissemination of new climate change knowledge to help communities adapt to the new environmental conditions. Such a structure also has the mission to translate the results of the Intergovernmental Panel on Climate Change's (IPCC) studies into a nationally digestible format so that these studies and the solutions IPCC proposes provide a basis for climate change discussion in the DRC. Doubtlessly, the bulk of knowledge produced by IPCC forms, eventually without much contextualization of some key parts, the largest bulk of the background knowledge to support any further research and back national climate actions.

Financing Climate Change via Sustainable Development Plans

Sorting out the climate change question requires large-scale action. It would imply changing current human lifestyles, which likens to reduce greenhouse gas (GHG) emission rates. To deploy large-scale climate action, we should consider the reality that funds are limited. Financing climate change action is not an easy task for the DRC and other less industrialized countries. Industrialized countries so far only partially kept their promise to disburse yearly committed funds for poor countries (Chapter 2). It is highly unlikely that industrialized countries will deliver on their promise by the 2025 deadline, even though reports were optimistic in the prospects for achieving the 2023 annual goals. Pessimism over Voïta's optimistic projections is plausible because only 6% of rich countries' contribution is truly new climate funding; the remaining of the climate-marked fund currently comes from old commitments toward sustainable development in the global South.[15]

Perspectives to increase new additional climate change funds are meager; the DRC should assign to its climate structure another strategic objective to help identify and mobilize internal resources. Levying internal funding is the first funding mechanism; external funds to support the national climate action should be counted on only secondarily. External climate funds are necessary, but they are only one way to finance national climate action. Climate national policies are an integral part of sustainable development public policies; they cannot entirely depend on international aid. The greatest funding source for climate action is to integrate climate change costs in national sustainable development budgets. A weighted arrangement of public and private structures around climate change requires becoming innovative in creating a financial architecture whereby each national financial asset incorporates the climate change burden in actions directed to reach the goal of the country's sustainable development.

By osmosis, imbedding financial assets in sustainable development blueprint allows all the national financial and other resources to influence all the desired action for sustainable development. To speak of climate-smart budgeting is not poetic. Budgeting climate-smartly is a process to map out national budgets factoring the needs to buffer human communities and ecosystems against the climate change hardships and vulnerabilities in the entire national life. At first sight, climate-conditioned and osmotic budgets seem intellectually senseless in a poor country like the DRC. Nonetheless, the beauty of this reflection is that poor countries remain structurally entirely to build. For the DRC, nothing would be lost to build its infrastructures with new climate and ecological standards. The DRC can spearhead climate change combat. The DRC pioneered the creation of protected areas in Africa,[16] which initiated the current scramble for protected areas in Africa. It is possible that the DRC sets the tone for the climate change issue. This is conceivably possible, provided that competent and conscientious people pilot the DRC's action; the DRC can leave an indelible mark because it has possibilities to do so. The question of funding is often a prioritization exercise and how to avoid leaking funds to inappropriate and illegitimate hands. The DRC can conceivably bear the costs of its own sustainable development and its adaptation to climate change, provided that it goes beyond the narrative of being a country that offers *the* solution to the climate problem without a clear agenda.

In addition to reasons discussed in Chapter 2, this chapter adds that applying COP's climate resolutions fails in the DRC because the fundamental climate questions are discussed using a scientific paradigm that does not speak to and is not rooted in African realities. This is the case because there is no appropriate structure to federate all the climate actions flowing from all national life sectors. Lacking a climate structure, climate actions are incoherent and fragmented to instill impactful changes. This chapter argues for the creation of a climate structure that supports research and the creation of DRC-contextualized climate knowledge. To fund this search and the creation of DRC-grounded knowledge, the DRC has internal financial means to support their own efforts, which can come from integrating the climate change question in the scheme and process of sustainable development. The DRC remains entirely to be built, as it is still a very poor country; climate change offers the opportunities to think innovatively and construct a new type of development. The chapter suggests that creating a climate structure is one way forward not only at the level of the DRC but also at continental level because African regions face similar climate realities. Surely, creating a climate structure has its own flaws; one of these flaws is the fact that despite its obvious merits that are laid out above, it does, in many respects, even seem to follow the trend of creating more and more structures, with the consequences of being inefficient. That is why, if created, such a structure should not be considered sufficient in itself; there is a need for more concrete action on ground. This is what Chapter 4 develops.

Notes

1 McCarthy N, Meirowitz A (2007) Political Game Theory: An Introduction. Cambridge University Press, p. 446.
2 Tshimanga RM, N'kaya GDM, Alsdorf D (Editors) (2022) Congo Basin Hydrology, Climate, and Biogeochemistry: A Foundation for the Future, American Geophysical Union, p. 592.
3 Mauss M (2012) Essai sur le don : Forme et raison de l'échange dans les sociétés archaïques. Présentation de Florence Weber, Presse Universitaire de France, p. 250.
4 Cai M, Caskey GW, Cowen N, Murtazashvili I, Murtazashvili JB, Salahodjaev R (2022) Individualism, economic freedom, and charitable giving. Journal of Economic Behavior & Organization 200: 868–884.
5 Arthur J (1993) World hunger and moral obligation: The case against Singer. In Sommers CH, Sommers FT (Editors) Vice & Virtue in Everyday Life: Introductory Readings in Ethics (Third edition), Harcourt Brace Jovanovich College Publishers, pp. 845–852.
6 Singer P (2016) Famine, Affluence, and Morality (with a foreword by Bill and Melinda Gates and a new preface and two extra essays by Singer) Oxford University Press, p. 120.
7 Sandel MJ (2020) The Tyranny of Merit: What Becomes of the Common Good? Farrar, Straus and Giroux, p. 288.
8 Marcantonio R, Javeline D, Field S, Fuentes A (2021) Global distribution and co-incidence of pollution, climate impacts, and health risk in the Anthropocene. PLoS ONE 16(7): e0254060. https://doi.org/10.1371/journal.pone.0254060.
9 Inogwabini BI, Matungila B, Mbende L, Mbenze AV, Tshimanga WT (2007) Great apes in the Lake Tumba landscape, Democratic Republic of Congo: newly described populations. Oryx 41 (4): 532–538.
10 Colom A, Bakanza A, Mundeka J, Hamza T, Ntumbandzondo B (2006) The socio-economic dimensions of the management of biological resources, in the Lac Télé – Lac Tumba Landscape, DRC Segment: A segment-wide baseline socio-economic study's report. Submitted to WWF DRC, p. 124.
11 Milton AD (2015) Forest resilience for livelihoods and ecosystem services. A dissertation submitted in partial fulfilment of the requirements for the degree of Doctor of Philosophy at George Mason University, p. 242.

12 Spencer J (1963) Language in Africa. Cambridge University Press, p. 167.

13 Delphi Messinger is the author of the 393-page Grains of Golden Sand: Adventures in War-torn Africa, published by The Fine Print Press in 2006.

14 https://www.unsdsn.org/.

15 Hattle A, Nordbo J (2022) That's Not New Money: Assessing How Much Public Climate Finance Has Been New and Additional to Support for Development. CARE Denmark, p. 61.

16 Inogwabini BI (2014) Conserving biodiversity in the Democratic Republic of Congo: A brief history, current trends and insights for the future. PARKS 20 (2): 101–110.

4 Climate Change

Adaptation and Mitigation in Central Africa and DRC

Introduction

The current public climate change discussion revolves around adaptation to the effects and mitigation of climate change. Current debates seem to oppose the proponents of adaptation as such and those who think that efforts should focus on mitigation. Dichotomy as it is the case in this preceding sentence takes people away from the common climate change evidence. Africans already experience climate change effects. For them, the question is not to separate adaptation from mitigation but to carry out activities on both simultaneously. Humanity must save what it needs to protect from what is already affected. Saving what can be saved should be done while humanity works to stop the degradation from worsening. To achieve these two parallel goals, this chapter argues that the citizens of Central Africa have no luxury to completely deviate from the global sustainability framework; anchoring their sustainable livelihoods on the global sustainability matrices is what they need. However, anchoring sustainable livelihoods on the global matrices of sustainability requires contextualization that makes sense for African culture and the civilization that the citizens of Central Africa want to give themselves.

This chapter provides the minimum conditions that can ensure that mitigation continues to apply concurrently with efforts to adapt to climate change effects. Physically, climate change is due to the imprisonment of infrared rays by the cap formed by greenhouse gases (GHGs) in the atmosphere. A disproportionately larger part of GHG emissions is man-made. Answers to these fundamental questions underlying the visible climate change effects are to work on this human-induced chuck of GHG emissions. African intellectuals must take charge of this subject to enlighten their compatriots. Without claiming to exhaust the subject, this chapter launches an intellectual debate on the possible paths toward progress that combines ecological requirements and social demands. With the proposition that African intellectuals must enlighten Africans being at the core of this chapter, the chapter also aims to loosely design a cookbook-type argument of how the global sustainability framework could be contextualized to fit the Congolese (Central African) conditions.

DOI: 10.4324/9781003493754-5

The chapter follows the logic that in order to sort out societal problems, identifying root causes of issues is the first step. With causes identified, solutions can be proposed. The difference between the argument in this chapter and other climate change scientific publications is that the chapter doesn't dwell on geo-chemical and physical causes. The geo-chemical and physical causes are significantly, lengthily and adequately addressed in different books and journals and are taken in this chapter as if they were already well-known and granted. Of course, stating that geo-chemical-physical climate change causes are given more than a sufficient space in climate discussions is not to infer that they are fully grasped by the broader human community. In many cases, there is still a big need for climate literacy, as the examples given in Chapter 3 described. Communities in Central Africa are much less informed of how humanity has come to where it currently stands with regard to the world climate. While history informs humans of how the civilization they built over centuries led to the current state of the world, the link between the current human civilization and the climate change remains much less firmly established in commonly shared ideas. That is why this chapter takes a different approach and discusses only the human side of climate change causes.

Climate Change's Causes Are Also Ethical

Climate change stems from *hubris*! Hubris is greed,[1] selfishness and human avarice. Ferry and Capelier[1] argue that hubris pushed human impact on nature to go beyond sustainable limits. Hubris leads to the five following problems[2] to which humanity must find solutions to maintain life on the earth: (1) the consecration of the individual, (2) the primacy of quantity over quality, (3) the cult of hedonism, (4) the dissolution of the sense of responsibility toward the community and nature and (5) the urgency of the present moment. Additionally, it is fair to add the idolatry of the material.[3] For Congolese (citizens of Central Africa), fighting hubris and founding a new civilization mean to find new ways for living a life whose foundations are the antipodes of the six abovementioned fundamental attributes of the current universal civilization. Paragraphs that follow go through each of the six problems and identify ways in which the problems arising from each can be addressed.

First, the consecration of the individual is a double-sided question. There is the place of individual vis-à-vis the society. The modern and globalized culture places the individual at the center of the social universe and makes each individual the ultimate goal of life. The Sartrean[4] existentialist philosophy, with the shift toward sovereign individuals, offered an easy flank on which the modern individual idea was built. In perspectives where individuals are sovereign, being independent is the supreme value; freedom becomes the ultimate goal that individuals pursue. Hence, each individual reclaims the full right to be the actor of their own life without suffering coercion from the world around them.[5] Being individual centered postulates that individual action can only be limited by the sky; the limits to individual human action do not really exist.[6] The limitlessness of human actions would remain as long as the individual does not encroach on the freedoms of others. '[...] Man is free, man is freedom', Sartre wrote.[7] Life where individual freedom is pushed to the extreme (individualism) leads to unbridled competition around material goods. Intemperate competition is the second modernism's trait. Competition is not inherently a bad thing; it may be a good thing when its aims are to release positive energies to construct a social order which strives for collective and individual success. Unfortunately, modern competition is rooted in social Darwinism, which applies the biological principle of the survival of the fittest to a social context where it has nothing to do. This led not only to the consecration of the individual but also, above all, to the idolatry of the material goods.[2] What boggles decently thinking minds is that most people currently confuse material goods with human prosperity and happiness. Thus orchestrated, the coronation of the individual dissolved the sense of responsibility to the community. The ordinary consumption patterns become the essential measures of prosperity and modernity's attributes. This is how the entire humanity has slipped into consumerism.[8] Despite differences in consumerism degrees across the world, there is now no part of the world where the drive to acquire unnecessary goods, gadgets and sometimes unwanted services to assuage abnormal human desires to show up social statuses is absent.

Originally from Western Europe and the United States, consumerism spreads across the entire globe to the point that humanity faces global consumerism.[8] The promise to extend consumption implies to munch increasingly more material goods and to demand still more and more new material possessions.[8] Insatiate demand for more new material possessions is the most powerful ideology of the 21st century; it intoxicates entire masses and promotes thoughtless capitalism with its religion of growth.[9] Surely, you cannot grow in a vacuum; there is a need of hardware to support the growth.[10] This is how the consecration of the individual dissolves any sense of responsibility toward nature. People draw everything that can be drawn from nature to consume all day long. More and more material is good and sound without even considering the quality in many instances. Surely, techniques to decouple economic growth and nature are emerging at micro levels[11] and should be able to sustain people to live decently. However, the emergence of these techniques should not prevent people to think about ecologically less onerous ways to live their lives.[12]

Individualism imposes superfluous desires on humans. Ironically, however, rather than singling out individuals, individualism, when combined with consumerism, equalizes everyone to appear like everyone else. A consumerist life merges ideas and goods to the point that wishes humans have are bundled to fit all at once. Humans are psychologically aggregated to have one taste to acquire still more material goods, possibly the same brands and qualities. As quantity takes precedence over quality, opulence betrays[13] its own foundations. Food, for example, instead of being a source of good health, the obsession with quantity leads to junk food which makes humans sick.[14] Excessive food production releases large CO_2 quantities into the atmosphere. Instead of producing sturdy goods to last longer, industrialists plan obsolescence which forces buyers to get rid of everything after a short time. Planned obsolescence, without recycling, has negative externalities, including resources depletion, pollution and weakened social protection. Absurdly, production's negative externalities are often neglected in the production costs' accounting. Negative externalities, unaccounted for, reduce everyone's symbolic heritage and discount the value of the natural capital's loss. Consumption ideology is simply unsustainable at its current levels; it leads humans to be at war against everything: against the biosphere, against their fellow humans and against themselves.[15] This metaphor may be too strong to make the point, but it is true that the current consumption levels are unbearable if no other alternatives are identified and used. Technologies promise to provide just such alternatives,[16] but they do not come at cheaper prices either! Hence, we are taken aback to the issue of the cost for humanity to survive through the current global crisis.

Fighting the biosphere, the individuals collectively puncture more resources from nature. Unfortunately, renewable and non-renewable natural resources are finite. The continued extraction of more from nature is to satisfy human sometimes useless and insatiable needs. Individuals wage a merciless war against others to monopolize the best shares of the common wealth. Cupidity drives the 1%[17] minority to accumulate more while pushing more than 35–45% of the humanity at the margins of social fragility. The wealthiest individuals find themselves at war with themselves because they are never happy; they constantly create more and more new desires and will never be completely satisfied.[18] Understandably, more consumption does not lead to the fulfillment of human satisfaction. Consumption is an auto-catalytic process that accelerates at a faster pace as it feeds itself.[19] Being an auto-catalytic process, consumption will always appeal for more consumption.

It is an understatement to admit that the consecration of the individual leads ultimately to the lack of responsibility. Humans have lost their sense of responsibility toward the community, nature and, inwardly, humans. This is plausible since the current blind capitalism civilization is built and blooms on the prepotency of consumption. Consumption levels cut human throats both figuratively and literally; in materially opulent countries, murder and suicide rates so eloquently speak about this situation. The situation is more complex because any consumer action is ecologically priced. CO_2 is one ecological price; it is world-widely spoken of now because its atmospheric content has become more harmful than normal. Over the entire 20th century, blind capitalism took over diverse life's modes and world-widely spread unstoppably. Its first dogma was that success meant to acquire more hardware, more electronic gadgets constantly remodeled and renewed, and to run faster. Fabricating all these gadgets to assuage human desires needs CO_2-hungry technical processes; succeeding in the thoughtless and blind capitalism era meant putting more CO_2 back into the atmosphere. To succeed today is to heat the biological matrices of life beyond the temperatures they can actually bear.

To get out of this situation in which greed and excessive selfishness plunged humanity, the first step to take is at the individual scale. Humans must leave their own current lifestyles. This step is about living a decent life without necessarily having useless gadgets and eliminating the most eccentric desires that unnecessarily torment their lives. This amounts to disconnecting from the current train of affairs and going slower. This is the hardest thing the current world, psychologically drugged by speed and instantaneity, will accept. The outcry aroused by the tax for the ecological transition in France, for example, confirmed this difficulty. Claims by Yellow Vests[16,20] were legitimate because the ecological transition taxes were to hit the most fragile of the capitalist system than the well-to-do guys; however, paying the price to de-carbonize the economy remains a non-optional path to save the earth from climate change disasters.

The second step to take is to get out of the instantaneity and the speed in which humanity is kept in hostage. The instantaneity trap means that humans no longer have enough time to think to make the rational decisions to guide their actions. Humans have to give themselves a three-level time space. First in thinking, humans should get out of the immediacy of action and inscribe the hopes and the vision of their destiny in a sufficiently long time. Paul Ricœur once asked humans to avoid the horizons of their expectations to land on the horizon of human collective experience.[21] Second, materially, to set humans free of the dictate of the present instant is a demand to create more durable and more qualitative things. Pushing back on the present instant's pressure is to reject planned obsolescence and the ever-continuous updates that make everything obsolete almost the moment it is exposed in a market. Continuous updates force people to immediately think of the next generation of objects immediately after purchasing them. Slowing down continuous updates doesn't suggest that change is a bad thing. To slow the pace is to say that change must resonate with the real human needs' rhythm; change should not instigate objectionable desires. Rethinking the present moment's urgency is to project humanity on the long temporal horizon. Placing one's life horizon in the humanity's long future temporal horizon is intergenerational justice. Intergenerational justice requires present humans to leave the earth to their descendants, if not in better conditions, at least as good and with sufficient materials as they have found it.[22] The intergenerational justice is hotly debated, raising questions such as whether present people were sure if future generations would really exist. If the future generations really exist, who they will be and how would they like to live their lives? These questions are raised and discussed from a Western-centered epistemology. The African ontology, for example, is constructed on the idea of a past-present-future living network. Africans hold that their ancestors are still alive,[23] and the present generation thrives to procreate for the future. Procreation is valued not for the immediate benefits but for the future of one's group. The continuation of the group is the reason for living people to ensure that their ascendants' blood is transmitted to the future. In this context, there are no questions about who the futures generations will be, what they would look like and what their needs would be. For Africans, the dead are not dead; leaving a healthy planet behind does not mean that one is leaving good living conditions to future generations only but also to one's own future life because one will never totally die. For cultures where rebirth and transmigration[24] prevail, the same questions about the future generations are also nearly irrelevant too. People who believe that their souls will depart anew for life after biological death, albeit in a novel forms (human, animal and spiritual),[25] see themselves as part of the future generations and aspire to leave the world in good conditions because, ultimately, they will re-use it. Many people who believe in reincarnation trust that the novel forms that they will have in the future are conditional to the moral quality of their actions in the present life or in the one that preceded our current time.[26] Responsibilities to future generations are embedded in the ways in which people perceive their lives. In an ontological environment where the present cannot be dissociated of the past and the future, the responsibility to each of the points of this triad past-present-future is compulsory for everyone. When one thinks one will continue to live, in one form or another, the responsibility for the future generations is not a generous act dictated by religious or moral obligations; it is part of the responsibility to one's self and can be seen as an investment for one's future.

The second deeper cause is the place of humans in nature. Basically, the place humans occupy in naturel things is synthesized in the dualism between nature and nurture. The long nature-nurture debate opposes two conceptual tendencies with two extremes. First, biocentrism says humans are part of nature and nothing sets them apart from other species. Humans are an integral part of the ecosystems in which they deploy their activities and lives. Biocentrism or naturalism implies humans should not be given more attention than any other species receives. Pushed too far, biocentrism leads to ecological monism whose consequence is that all species come strictly from the same origin. Just as other living beings, humans derive life from the single source, which leads to isonomy in nature. Biocentrism has its supporters in the circles such as deep ecology (Chapters 5 and 6). Deep ecology sometimes takes the religious forms[27] and often preaches the thin human importance vis-à-vis the nature.[16] Second, anthropocentrism accords to humans a special place in nature; it says that humans have little traits in common with other species in nature. For hardline anthropocentric people, humans must master the earth. For anthropocentrism, humans are the measure of all success. Anthropocentrism's supporters are among those who have a limited reading of the Holy Scriptures (Bible); they claim to draw their reason from Genesis according to which God is said to tell Adam and Eve to be fruitful and prolific, to fill the earth, to dominate it and subdue the fish of the sea, the birds of the sky and every beast that stirs on the ground. Anthropocentric extremists don't care about the more nuanced version that the men-command-all reading of the Bible is but partial. Indeed, without forcing the exegetical lines, the same Genesis imposes a limit on what humans can do with nature; humans were given the permission to eat fruits of any tree in the garden except for the tree of knowledge. Excluding the fruit of knowledge from being eaten limits humans in what they can extract from nature as Jack Miles wrote[28] and for whom there are two accounts of creation. In the first account, something is commanded, but nothing is forbidden, and in the second one, prohibition is slipped in, and it is imposed in the interest of humans.

But of what nature do deep ecology and ecological anthropocentrism talk about? The nature-nurture debate has led to the emergence of two concepts of nature: *Natura naturata and Natura naturans*. It depends on how humans understand these two concepts of nature that they formulate a world view on which they decide how to treat nature. It based on how humans understand these concepts that they position humans in nature. *Natura naturata* is *natured* nature and is always defined as created nature or nature-made nature. *Natura naturata* is the idea that nature is an object in human hands. *Natura naturata* implies that humans possess nature; they can dispose of it as they wish. *Natura naturata* founds anthropocentrism. *Natura naturans* is given nature (*naturing* nature); *Natura naturans* is nature in the self-creation process. *Natura naturans* is a nature that manages vital nature-object flows. Things are more complex than they are herein stated; as such, they shouldn't be seen as black spots on white boards. These broad strokes (unsophisticated as they are) give the general picture of what core messages *Natura naturata* and *Natura naturans* convey.

Nature is a complex but highly organized and stable whole; nature is crossed by flows of energy.[29] Stating this seems to endorse those hastily reach clear-cut conclusions in favor of *Natura naturans*. But the place humans occupy in nature is a complex subject. Within the nature-culture dualism, humans are not totally confounded with nature, and they aren't essentially cultural creatures.[30] Humans should be viewed always as both nature and culture. Nature gives humans the existential biological foundations while at the same time, humans use nature to build culture. The two human dimensions seem inseparable. A message this bi-dimensionality sends is that humans should not turn themselves into destroyers of nature, which is the foundation of life. Nature provides material benefits to human cultures. Cultures create in humans needs dictated by human instincts to cling to life. Some desires are solely cultural constructs and could be ignored. Culturally formatted, humans can no longer remain passive actors within nature. Passivity is ceased in human cultures because natural chauvinistic instincts and culturally created insatiable desires force humans to continue with creation, one way or another. Bludgeoning of natural chauvinistic instincts and culturally idealized insatiable desires hook humans to do more to gain still more.

Admitting the human bi-dimensionality points toward some generic solutions to the problems posed by climate change may be identified, tested and, possibly, implemented with some measures of success. Without fully endorsing Aldo Leopold's ideas, his proposal, according to which human action on nature would only be fair if it maintained the stability, integrity and beauty of the various biotic sets in the world,[31] re-poses questions of ecological thresholds and the need to conduct ecological cost-benefit analyses of human actions. An action with negative ecological balance sheet is not fair even if its economic repercussions are fivefold the financial investment. Negative ecological balance means to include both ecological and social externalities.

The third cause of climate change and the biodiversity crisis is the rampant increases in human demography. Since the beginning of the industrial era, the earth's human population has been growing significantly. Projections are that the earth's human populations will grow from 8 billion people (2023) to 8.5 billion (2030),[32] an increase of >5%. Africa is projected to go from 1.25 billion Africans (2023) to 1.7 billion in 2030 (an increase of 36%). The Democratic Republic of Congo (DRC) is to move from 82 million people (2023) to 120 million (2030) and even bleaker to 197 million people (2050). The DRC outlook is even more alarming; it represents increases of 48% and 143%, respectively, compared to 2023. By looking at such increases in human populations in a culturally fashioned-consumerist world, it is difficult to see how humans will stop climate change, whatever investments they make. Natural resources and the world in which humans live are physically finite! Solutions to climate change require a better planning of the usages of natural resources and of numbers of humans. Certainly, it is no question to stop families from procreating; it is more a matter of envisaging a population growth that is reasonable and sustainable by a planet whose available resources dry up as humans occupy all of its parts. Humans currently occupy even parts of the world that once served as refuges for other forms of life.

Perspectives on Possible Solutions for the Congo Basin

The generic climate solutions framework contains the following elements: (1) getting rid of one's part of current lifestyle, (2) escaping the trap of immediacy, (3) maintaining ecological stability, integrity and beauty and (4) envisioning population growths that earth can sustain. This generic framework can be contextualized for the DRC. Everything said on the DRC applies for other Congo Basin countries, with contextual adjustments.

First, Congolese must reinvent the meaning of solidarity. This means that Congolese should ensure that everyone participates, within the limits of their proven capacities, in maintaining the social safety net which supports the upward social progress[33] for all. Working on a social safety net doesn't have to go back to the Marxism-Leninism's creed, as some neoliberals[34] would think it to be. Upward social progress for all is a difficult and lengthy task, but it isn't impossible. Upward social progress for all is possible as Scandinavian countries show. Surely, the Scandinavian model will need some major readjustments to operate in other contexts. Without aping the Scandinavian countries, Congolese can dust Congolese solidarity of its traditional heaviness that opposes individual efforts and progress; the DRC can contextualize the social fabric model of the Scandinavian countries to fit the DRC's situation. Congolese, for example, can proportionally pool together the means for an environmental education that emphasizes the importance of people. Congolese can also pool the means for healthier and common means of transport instead of suffocating their countries by a senseless rush toward obsolete and increasingly polluting vehicles and transport themselves otherwise.

Second, Congolese must re-enchant life in villages.[35] This will depollute enormously by reducing human concentrations in the slums around large cities. Re-enchanting the village means bringing security, health, education, electricity, transport and communication to villages. Investing in these areas gives meaning to agriculture without which sustainable development doesn't exist. Prosperity resulting from agriculture, the DRC can de-carbonize its economy by reducing agricultural industrialization and long-distance transport of food from production sites to markets. Re-enchanting the village is about creating more small- and medium-size towns to release the pressure over nature.

Third, Congolese must invest in the long-term thinking and the long time. Investing in the long-term thinking for a long time amounts to taking Congolese out of the lure of easy and immediate gain. Short-term gains make Congolese think in the mode of amaranth cultivation. Amaranths give results in a relatively short time; Congolese prefer to cultivate more amaranths instead of orchards in to gain quickly. The lure of immediate gain's culture means that Congolese think less about the consequences of their actions to the point that they can even import Western waste that comes to spoil their life in their countries. For example, vehicles prohibited from circulation in the Western countries over-abound in the DRC. Imported obsolete vehicles pollute the atmosphere to proportions which have yet to be measured but which are certainly above authorized thresholds. Investing in long-term thinking is nothing more than breaking with chronic improvisation,[36] which keeps Congolese resourceful yet does not lend itself to a thinking process that can take the current and future issues into account because everything is improvised. Improvisation is an attribute of short-termism.

Fourth, Congolese should break the propensity of the 'copy and paste' culture, which makes Congolese think that the life model is unique and universal.

Congolese think that development is the efforts to mimicry anything that is American or Western European without examination. Breaking 'copy and paste' culture isn't to reject the idea of learning from experiences of others. Certainly, it is foolish to reinvent the wheel all the time throughout human history. Reinventing the wheel is bad because it means uselessly consuming too much time. Reinventing the wheel also denies the contributions Africans have made to the global human civilization. However, there is no one-fit-all economic model; Congolese must invent their own economic model. The DRC doesn't have an economic model yet; when the DRC invents its economic model, it shouldn't be American, Rwandan or Brazilian.[37] The American model is not what is needed in the DRC conditions, the American economic model being essentially consumerist to the point that no one outside of the US context wants to import it untouched. The so-called Rwandan economic model doesn't yet really exist and still needs to prove itself before talking about it so much and before it is sold to other African countries with different geographical and cultural conditions. Let us not forget so easily that Africans experienced such dithyrambic ecstasies in the recent past. The Ivory Coast of Houphouët Boigny, the Zaire of Mobutu, the Uganda of the early-day's Museveni, for example, were, for distinct short periods of time, models that the West was frenzied about to sell their economic potions to other developing nations.

The Brazilian model being ecologically and socially expensive,[38] the model that matters is the one that is economically viable, socially equitable and ecologically sound. These three pillars will make the DRC meet the requirements of development, demands of social equity and economic de-carbonation. These pillars can work only if applied without falling into sustainable development imposture traps, which consists of dressing up obsolete development concepts with new words[39] and rebranding the old aid budget to espouse new jargons (Chapter 2). One task of Congolese intellectuals isn't only to swallow experiences from elsewhere but to build and test tools to reflect and act within their country-specific ecological, geographical and historical conditions. This isn't an isolationism's sermon; Congolese intellectuals can only be usefully open to external contributions if they have a better understanding of who diverse Congolese are, what they lack in their local geographies and what they want under these circumstances.

Fifth, the DRC should know that its current population growth negates any effort it invests in sustainable development. Dense population isn't a problem in itself; DRC needs enough brains and enough expert hands to ensure its optimal sustainable development. Sufficient and well-educated populations are a critical economic asset. But population growth is an economic asset only when it is within the capacity the country can carry. The current trends in DRC are that human population grows in excess of the country's economic growth and above the country's capacity to carry it. Looking at new numbers of human heads entering life in DRC suggests that the situation rapidly gets beyond manageable limits. DRC's enormous natural resources and ecological diversity are the country's strength and can be used to cope with this trend. Unfortunately, population growth calls for more quality education for the youth to be adequately accommodated. For this, DRC needs human resources capable of producing wealth, but producing affluences needs to be matched with the country's natural limits. Under current conditions, a population growth greater than economic growth rate in DRC increases the carbon-dependent economic growth because the country lacks research capacities, and its economy is carbon-energy dependent; population growth will amplify the energy demand. It is illusory to talk about adaptation and mitigation without thinking of the human population numbers the earth can sustainably contain.

Finally, national land use planning should be rethought and implemented throughout DRC as both climate adapting and mitigating strategies. As a means to adapt to the climate change effects, an intelligent national territory planning is useful for people's lives in areas that are less vulnerable to climate change effects. As a mitigation tool, the skillful planning of the national territory will serve as a mechanism to save different climate functions through the allocation of ecological functions to each square kilometer of each of the national territory. Attributing functional usage to each land piece should be done to avoid harming fundamental climatic balances. This skillful arrangement will allow DRC to organize hospitality for climate internally displaced people or refugees. Climate change effects already displace and will continue to displace Africans. The DRC will be, with their extent and their ecological potentials, land of African immigration. Rethinking land use planning differently from what is currently will make it possible to absorb migratory flows without having to lose sight of the fact that each country can only bear part of this African burden and that no country will be able to bear the entire drain on its own.

Notes

1 Ferry L, Capelier C (2014) La plus belle histoire de la philosophie. Robert Laffont, p. 456.
2 Sarr F (2016) Afrotopia. Philippe Rey, p. 155.
3 Cox H (2016) The Market as God. Harvard University Press, pp. 307.
4 Vegleris E (2009) Vivre libre avec les existentialistes Sartre, Camus, Beauvoir... et les autres. Eyrolles, pp. 139.
5 Ennuyer B (2016) Individu et société : le lien social en question? ' Ethics, Medecine and Public Health 2: 574–583.
6 Fontaine P (2008) Sartre: penseur de la liberté. Conférence-débat avec Philippe FONTAINE organisée par Professeur Czeslaw Michalewski à Sèvres, le 14 octobre 2008, à la Maison Pour Tous de Ville d'Avray.
7 Sartre JP (1996) L'existentialisme est un humanisme. Gallimard, p. 108.
8 De Munck J (2011) Les critiques du consumérisme. In Isabelle Cassiers et alii (Editeurs), Redéfinir la prospérité. Jalons pour un débat public. Editions de l'Aube: 101–126.

9 Campbell C (1987) The Romantic Ethic and the Spirit of Modern Consumerism. Blackwell, p. 294.

10 Inogwabini BI (2019) Ecology and sustainable development. In Leal FW (Editor) Encyclopaedia of Sustainability in Higher Education. Springer.

11 Henry C (2023) Pour éviter un crime écologique de masse. Odile Jacob, p. 239.

12 Levrel H, Missemer A (2023) L'Economie face a la nature : De la prédation a la coévolution. Les Petits Matins, p. 247.

13 Dupuy JP, Robert J (1976) La trahison de l'opulence. Presses Universitaires de France, p. 256.

14 Levenstein H (2003) Paradox of Plenty: A Social History of Eating in Modern America. University of California Press, p. 362.

15 Caye P (2015) Critique de la destruction créatrice. Les Belles Lettres, p. 336.

16 Ferry L (2021) Les Sept Ecologies : Pour une alternative au catastrophisme antimoderne. Editions de l'Observatoire/Humensis, p. 274.

17 Stiglitz J (2015) La grande fracture : les sociétés inégalitaires et ce que nous pouvons faire pour les changer (Traduit de l'Anglais par Françoise, Lise et Paul Chemla). Les Liens Qui Libèrent, p. 477.

18 Barthes R (2013) Toward a psycho-sociology of contemporary food consumption. In Counihan C, Van Esterik P (Editors) Food and culture: A Reader. Routledge, pp. 23–30.

19 Diamond J (2000) De l'inégalité parmi les sociétés (Traduit de l'Anglais par Dauzat PE) Nouveaux Horizons, p. 605.

20 Bate F (2018). France's Macron learns the hard way: Green taxes carry political risks. Reuters 2 December 2018.

21 Baroni R (2010) Ce que l'intrigue ajoute au temps Une relecture critique de Temps et récit de Paul Ricœur. Le Seuil Poétique 3 (163): 361–382.

22 Desjardins JR (2013) Environmental Ethics: An Introduction to Environmental Philosophy (5th Edition). Wadsworth, p. 179.

23 Tempels P (1969) Bantu Philosophy. Présence Africaine, p. 190.

24 Mills AC, Slobodin R (1994) Amerindian Rebirth: Reincarnation Belief among North American Indians and Inuit. University of Toronto Press, p. 410.

25 Rittelmeyer F (1988) Reincarnation: Philosophy, Religion, Ethics (Translated from German by Mitchell ML) Floris Books, pp 144.

26 Stevenson I (1996) Twenty Cases Suggestive of Reincarnation (2nd Edition). American Society for Psychical Research, p. 362.

27 Taylor B (2001) Earth and nature-based spirituality (Part I): From deep ecology to radical environmentalism. Religion 31 (2): 175–193.

28 Miles J (1996) Dieu : Une biographie (Traduit de l'Anglais par Dauzat PE). Robert Laffont, p. 455.

29 Leopold A (1949) A Sand County Almanac. Oxford University Press, Ballantine Books (Reprint edition, 1986), p. 295.

30 Bookchin M (1990) The Philosophy of Social Ecology: Essays on Dialectical Naturalism. Black Rose Books, p. 198.

31 Callicott JB (1989) In Defense of the Land Ethic: Essays in Environmental Philosophy. State University of New York Press, p. 336.

32 United Nations Department of Economic and Social Affairs (Population Division), World Population Prospects: The 2017 Revision, Key Findings and Advance Tables. United Nations, p. 46.

33 Hirsch J (1978) The state apparatus and social reproduction: Elements of a theory of the bourgeois state. In Holloway J, Picciotto S (Editors) State and Capital: A Marxist Debate. Edward Arnold, pp. 57–107.

34 Fischer AM (2020) The dark sides of social policy: From neoliberalism to resurgent right-wing populism. Development and Change 51 (2): 281–726.

35 Bookchin M (1995) Re-enchanting Humanity: A Defense of the Human Spirit against Antihumanism, Misanthropy, Mysticism, and Primitivism. The University of Michigan, pp. 343.

36 Mbembe A (2013) Sortir de la grande nuit, essai sur l'Afrique décolonisée. La Découverte, p. 254.

37 Kamerhe V (2011) Les Fondements de la politique transatlantique de la République Démocratique du Congo : La République Démocratique du Congo, terre d'espoir pour l'humanité. Larcier, p. 224.

38 Montoya MA, Allegretti G, Bertussi LAS, Talamini E (2023) Domestic and foreign decoupling of economic growth and water consumption and its driving factors in the Brazilian economy. Ecological Economics 206: 107737.

39 Latouche S (2003) L'imposture du développement durable ou les habits neufs du développement. Mondes en Développement 31/1(121): 23–30.

5 De-carbonating Developed Economies and the Right to Development

Introduction

Reducing greenhouse gas (GHG) emissions in the wealth creation process and sustain economies is now a pre-requisite for all new development initiatives. Reduced GHG emissions is fundamentally important to sustain life on the earth; wealth production has to be done while keeping emissions at acceptable thresholds. Most of human development activities release GHGs into the atmosphere and heat up the earth's temperature (Chapter 1). Heating up the atmosphere began in the 18th century and reached unprecedented levels by the end of the 20th century. Both saving life and maintaining economic prosperity need strategies to reduce GHG emissions. This is a truism, but recalling this basic truth will, from now onward, remain important to all and all the time. This chapter looks at GHG reduction mechanisms. Climate experts name GHG reduction mechanisms mitigation. This chapter doesn't list what should be done to limit GHG emissions; it analyzes how some proposals to mitigate the climate change are translatable in non-industrial contexts. Mitigation is defined as a set of human actions to keep GHG emissions at moderate levels that are within the limits of life's sustainability on the earth. Simply put, climate change mitigation equates to slowing down the release of GHG emissions.[1] This chapter discusses the implications of maintaining forest cover in Central Africa on forest-dwelling communities and how numerous proposed mechanisms at Conference of Parties (COPs) are felt in Central Africa. Particularly, the chapter discusses mechanisms such as forestation, reforestation and the idea of keeping forests intact to continue being safe sinks for world-widely emitted CO_2.

CO_2 Emissions per Country per Year

Before discussing why a good idea as de-carbonizing the world's economy is problematic for Central African countries, let's look at some facts. An obvious climate change fact is that all countries don't emit the same amounts of CO_2 yearly. In 2016, 10 CO_2 super emitter countries emitted 67.3% of the

DOI: 10.4324/9781003493754-6

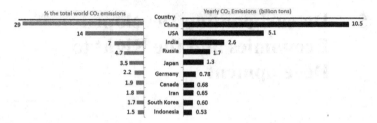

Figure 5.1 2016's CO_2 per super emitter countries and associated % emissions.

Source: www.worldometers

total world's emissions (Figure 5.1). The first five super emitter countries were China, the United States, India, Russia and Japan that emitted 58.2% of the world's CO_2. They were followed by Germany, Canada, Iran, South Korea and Indonesia with 9.1% of the total world's CO_2 emissions.

In 2023, the five CO_2 super emitters remained the same; Canada disappeared from the list, whereas Saudi Arabia was on the eighth row. Iran surpasses Germany, and Indonesia outdoes South Korea (listed 10th).[2] Without stigmatizing any country, once yearly per capita emissions are considered, ranks change as shown in Figure 5.2. Using the same 2016 database, China doesn't appear in the per capita yearly list, whereas the United States and Russia appear in the last rows. This fact should not acquit the three countries of their responsibilities as the first three CO_2 super emitters. In terms of year per capita emissions, smaller countries such as Qatar and Montenegro lead the queue.

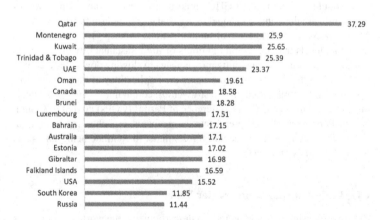

Figure 5.2 CO_2 Emissions per capita for countries emitting more than 10 tons per person.

Source: www.worldometers

That smaller countries like Qatar and Montenegro lead the queue in per capita yearly CO_2 emissions is not new knowledge. However, this shows why discussions about reducing emissions can be emotionally charged and less palatable for many countries. The most industrialized countries loosely comply with the agreements reached at COPs; they cannot plausibly explain this negative attitude. However, per capita emissions show why sometimes the most industrialized economies shift their ideas when CO_2 emissions are brought in per capita. The two world's biggest economies (the United States and China) contribute 43% of the CO_2 emitted globally, but each American and Chinese emit less CO_2 than each Qatari, Montenegrin, Luxembourgian and Gibraltarian. Qatar, Montenegro, Luxembourg and Gibraltar's economies are comparably smaller than those of the United States and China.

Whether one takes crude CO_2 emissions per country or one goes with the per capita CO_2, no African country is listed among the 10 CO_2 super emitters. South Africa, with its 391 million tons of CO_2 per year (per capita 7 CO_2 tons per year), emits the most CO_2 in Africa and is the 15th world's CO_2 emitter. The Congo Basin countries are not among the major CO_2 emitters (see Table 5.1). Congo Basin countries shelter most of the African forests; they are countries on which the international community exerts high pressure to keep these forests intact to buffer the climate change effects.

The same pattern as above emerges here too; countries emitting the most CO_2 are not necessarily the ones whose per capita emissions are the largest ones. Gabon has the highest CO_2 per capita in the Congo Basin, but it is the third emitter in total yearly CO_2 emissions. Conversely, the DRC is second in total yearly CO_2 emissions in the Congo Basin, but its CO_2 per capita emission is the lowest. These Congo Basin inter-country differences should not, however, hide the fact that the Congo Basin's combined 29 million CO_2 tons per year are not comparable with emissions of countries such as China, the United States or even South Africa. Taken prima facie, one understands, at least partially, why the people of Congo Basin are angered that they are the first to be asked to make sacrifices to protect the world against climate change without additional means to support their livelihoods (Chapter 2).

Table 5.1 CO_2 emissions per Congo Basin country and per capita CO_2 emissions

Country	CO_2 emissions/year	CO_2 tons/person
Cameroon	9.5 Million tons	0.4
DRC	6.6 Million tons	0.08
Gabon	5.7 Million tons	3
Republic of Congo	5.0 Million tons	1.1
Equatorial Guinea	2.0 Million tons	2
Central African Republic	543,000	0.1

Source: www.worldometers

Pathways to Mitigate CO_2 Emissions

There are several ways to mitigate CO_2 emissions; most are amenable to five critical actions that, if combined, offer different scenarios to mitigate CO_2 emissions. Climate change experts laid out combinations of scenarios and the options, which are mitigation strategies. Having several mitigation strategies implies that strategies aren't to be implemented singly and on a one-by-one approach. For mitigation strategies to work, it is critical that actions are implemented in parallel and simultaneously. The backbones of sensible mitigation strategies include (1) renewable energy sources,[3] (2) sustainable transport, (3) more energy-efficient buildings, (4) sustainable land uses and (5) maintaining forest covers[4] across the world.

The discussion that follows focuses on maintaining forest cover intact; some people urge to increase forests across the world. Before going into why the idea to keep the forest cover intact is controversial[5] for the DRC, it is essential to say something about why forests are important in mitigating the effects of climate change. The importance of the forest cover as part of the climate change mitigation strategy stems from the photosynthesis process. Photosynthesis is a natural process wherein a plant combines water it extracts from soil by its roots with the air's CO_2 the plant's green leaves capture to produce sugar using the solar rays from the sun (solar energy). Plants dissociate the air into CO_2 and oxygen; they release oxygen and absorb CO_2 in daylight time. Note that trees absorb CO_2 in daylight time but release it at night. CO_2 is the culprit and most conspicuous GHG. Forests absorb CO_2 and as such, they are also natural biological mechanisms humans have readily available to appropriately manage CO_2 to keep the world climate under control.[6] Forests are not the only natural mechanism that keeps the world's climate within the margins of heats that tolerate diverse life forms. Oceans and other large water bodies capture CO_2 from the atmosphere; soil too is a major sink in its own. But stating that forests are not the single CO_2-captor mechanism does not decrease the importance of better and sustainable management of forests as one of the most obvious and one of the most effective pathways to mitigate climate change.

To know why forests matter to sort out the mess humans created with the climate is good but up to a point. Forests have been capturing and will continue to capture CO_2 from the atmosphere and convert it into the sugar and greens that wildlife and humans live on. Complications emerge at the point of thinking about how much forests are left on the earth and where they are located. Most industrialized countries have used their forests to construct their economic development. Forests were part of the economic growth in today's world wealthiest nations.[7] This is where de-carbonating economies and the right to development for Africans clash (Chapter 2); they clash because schemes of de-carbonating production rely largely on the possibilities to keep the current forests intact to absorb the CO_2 economic activities emit world-widely. For Congolese, forests are daily livelihood's key resource and are so important for DRC's development. The request to keep forests intact sends the message that Congolese should indolently accept to remain under the yoke of poverty.

Greening the Economy

Greening economic activities is a simplified and jovial way to stipulate that economic activities should use processes that reduce quantities of CO_2 they emit. Surely, very scarce economic activities don't emit CO_2. Greening economic activities means reducing the CO_2 nearly all human activities emit. The mitigation mechanisms identified earlier in this chapter can achieve CO_2 emissions reduction. If everyone can make some proportionate efforts to reduce their CO_2 emissions for their productive activities, it will be good and should be supported by all means. However, greening economic activities became an issue once CO_2 was converted in a marketable good. The CO_2 trade is hard to understand when forests are its raw material. Putting a value on planted, restored and maintained forests is complex. Politically, when the United States introduced the CO_2 trading provisions in the Tokyo agreement in 1997, the process to green economy was turned into a right to continue emitting CO_2[8] for those who can buy rights pollute. This provision implied that those who can afford purchasing everything can continue to emit CO_2 and keep their lifestyles as they wish and unscathed. In many ways, trade in CO_2 is felt to be a meek semblance for the richest countries to reduce their own emissions. In reality, trade in CO_2 has introduced a mechanism through which the world's richest countries would bail their way out of any sincere commitments to reduce GHGs[9] world-widely.

CO_2 trade gives the right to powerful nations to continue doing their businesses as usually. Clearly, greening economic activities, using CO_2 trade, demands countries incapable of purchasing things to bear the blunt and heavy costs to reduce GHG emissions globally. Trade in CO_2 is about foresting, reforesting and maintaining the current forest covers intact across the world. The way the CO_2 trade emissions should work is that one rich country that isn't technically or unwillingly able to reduce its national CO_2 emissions looks for a country that can accept purchasing the rich country's emissions. Parties agree that the richest country will continue with its economic activities unabated, whereas the poor country will work to reduce its own emissions while planting new forests and reforesting degraded ones and accept to maintain its actual forest cover intact to absorb the CO_2 richest buyer country emits. Standing trees and forests become marketable raw materials that poor countries sell to the most industrialized world. If the only logic to tackle the world's most vexing issues was the market forces, CO_2 trade seems logical. Unfortunately, markets are not always the only way to solve some world's most troublesome issues. GHGs, including CO_2 emissions, are such issues for which market forces can even worsen the situation. Decades after 1997 reveal that climate change solutions were not in singly identifying and implementing GHGs workable market principles, as Americans thought. Many of the climate change issues, including trade in CO_2 emissions, are rather ethical, political and technical ones. Climate change issues, including the trade in CO_2, raise serious moral questions. Technically, quantifying the volume of CO_2 to trade remains a complex issue for most people in less technologically advanced countries. The technicalities of CO_2 quantification are also complex for most laypeople in the most advanced countries. The complexity of the techniques to measure the quantity of CO_2 to be sold creates a situation where the countries that sell the CO_2 sequestered by the forests they plant, the zones they reforest and forest covers they keep intact are dependent on the CO_2 buyer countries to help measure the quantities being traded. This is a weird state of affairs where the buyers alone know the amount of the good they are buying. The international CO_2 trade falls in the same old trap where powerful buyers establish raw materials' prices. The old patterns of the world economic divisions remain safe. The world center and the modern world's periphery keep their respective historical positions.[10]

Monitoring, verifying and reporting on GHG emissions themselves is another area where things go awry with CO_2 trade. CO_2 seller countries may trade CO_2 in some plots while conducting activities that emit equal or even more CO_2 in other plots, which is officially CO_2 leakage. How does one monitor, verify and report on such situations when tools are very complex to grasp by the laypersons and, sometimes even by some well-educated ones? How does that work in a situation where wealthier CO_2 buyer countries impose the rules and techniques to be used by CO_2 seller countries? How would that work in situations where national legislations on climate change are so weak in CO_2 seller countries? For CO_2 seller countries, environmental legal frameworks are lower priority than, for example, fighting generalized insecurity with limited means or feeding people throughout a single calendar year, etc.

Overlaying GHG-super emitter countries on the currently most forested countries makes complex facts even more puzzling. Currently most forested countries are where abject poverty levels highly prevail, whereas GHG-super emitters are the world's richest nations. Viewed this way, the CO_2 trade may imply that the richest countries should continue to develop, keeping their comfortable economic situations, and keep advancing. Conversely, CO_2 trade may also chiefly infer that the poorest countries, CO_2 sellers that keep the world's weather bearable, should remain in primitive life ways. Congolese often feel that poor countries are forced to accept primitive lifestyles and material poverty to support the steady-state growing opulence of the richest countries. This feeling explains why for Central African countries, right to development precedes any deal asking the people of Central Africa to reduce CO_2 emissions (Chapter 2). Foresting new areas, reforesting former forests and keeping the remaining forest covers intact should not equate to keep some of humanity's largest populations under the bondage of utter poverty. They argue that even if forests were the only way to salvage humanity, the way to deal with the situation should not repeat what happened throughout the world injustice history.

For Congolese, it was ironic to see a gas-swallowing jet carrying just one passenger (with some guards and secretaries) from Washington DC to Kinshasa to press Congolese to accept not using the forests to curve the difficult climate change situation humanity has been brought in. Crystal clearly, using this mind boggling example doesn't absolve Congolese from acting responsibly to contribute to the global efforts to mitigate GHG emissions. Congolese moral responsibilities come from the geographic position they occupy (Chapter 3); Congo Basin peoples have to be careful not to destroy forests and other ecosystems the way the current industrialized world has done. Congo Basin people can progress differently, which they can do if they increase their investments in research and innovations. That Congo Basin people can economically progress differently seems a wishful thinking, which is sensible because no Congo Basin country seriously invests in research and innovation. Congo Basin countries' research and innovation investments barely exist because of the paucity of resources they have. Even when some resources become available, Congo Basin countries' higher priorities are not research and innovation because the poverty's burden still nails down most peoples. With the example where the US special climate envoy asked Congolese to behave in climate sensitive ways,[11] if there is any sincerity in the Washington's quest, it should be in helping governments of the Congo Basin to invest in research and innovation[12] to help them to embrace new directions toward sustainable economic growth and prosperity. Asking these people not to cultivate their foods to avoid deforestation seems to be beyond irony and closer to an economic cynicism whereby the ones that emit more CO_2 ask those who already consume <3 kWh daily to limit their cooking energy to salvage the world.

Before the lesson-giving trip from Washington DC, we lightly saw the CO_2 leaking notion. CO_2 leaking goes beyond contexts of planting new forests, reforesting formerly degraded ones and keeping stretches intact forests continuously intact. Bailing themselves out of commitments to reduce their own CO_2 emissions, industrialized countries also leak a lot of their CO_2 via other pathways. Central African countries have become technological dumping sites for industrialized countries. Industrialized countries close most CO_2-emitting industries to reduce their emissions, but they also displace GHG sources to poor countries. Old gas-thirsty vehicles,[13] garbage and polluting products are officially exported and are cheaply sold to African countries.[14] While walking in streets of Kinshasa or Yaoundé, one is surprised by the numbers of American-made large and heavy four-wheel driving Chevrolets and Fords that grumble around. For the newly rich people in Central African cities, Chevrolets and Fords symbolize political and financial might. The politically and financially mighty people in Congo Basin import CO_2

sources in Central Africa. Central Africa's politicians should not only brawl against what Western industrialized nations do against climate. Newly rich people in the Congo Basin should also quell down and bring under control their own thirst to demonstrate their power through material goods. Asking politicians of Central Africa to avoid brawling against Western industrialized nations and asking them also to keep their own emissions under control don't decrease the importance the right of the citizens of Central Africa to aspire to some levels of decency and higher stages of genuine sustainable development deserves. For the development in question to be sustainable, Central Africa's citizens should be empowered to produce their own food to reduce their CO_2 emissions; the CO_2 footprint of Central Africa's citizens significantly increases when the energy taken to produce the food they eat and the energy to transport it to reach the consumption sites are added together. Most food people eat in Central Africa is produced in industrialized countries, which means it travels from afar before it reaches where it is to be consumed. Central Africa's communities' per capita CO_2 emissions are often minored by omitting CO_2 emissions to produce the food people eat and CO_2 emitted to transport the food along long distances to reach points of its consumption. There are historical, political and structural reasons why the production of food by local communities in Central Africa has been lagging behind. These reasons include environmental factors such as limited soil productivity and international policies such as food market contortions subsidies in Western agricultural industry cause. It is beyond the scope of this chapter to discuss the reasons why African agriculture is incapable of feeding Africans in Africa. However, to mention just one fact that is politically unjust and could be dealt with if politicians in the developed countries had the will to change things. The subsidy policies in Western countries simply kill any local food production efforts of Congo Basin countries. For example, when the chicken produced in the European Union's countries is transported to the DRC and sold at costs that are 3 times less than the chicken produced in the DRC, the contortions of the food market are obvious. Ironically, industrialized countries give subsides to their agriculture while simultaneously providing budget supplements to Congo Basin countries; budgetary supplements help poor countries import goods. Policies to protect the agro-food industry of Western developed countries have their own impacts on climate change. The public policies to protect agriculture in developed countries increase per capita CO_2 emissions not only in their own countries but also in the less industrialized ones. The Western agro-food industry's protective policies contribute significantly to leaking CO_2 emissions to countries where the Western-produced food is consumed. Food production in Central Africa just shows how the issue of climate change is tricky to deal with.

Theodor Adorno thought that simplicity was holy (*Sancta simplicitas*)[15]; this would logically conclude the discussions on *hubris*. *Sancta simplicitas* should be nuanced in view of Central Africa's life conditions. The goal of the efforts to diffuse climate change should remain to have a healthy earth where necessary conditions to maintain material bases of life are kept safe. The idea *Sancta simplicitas* holds all of it values in that process; however, it should be also re-qualified. Integrating the right to development was an essential demand from Central Africa to begin efforts to mitigate climate change (Chapter 2). What is needed is a little bit of more material wealth for communities languishing under the burden of poverty, not simplicity in life as such. In the DRC, poverty is a climate change driver. For city-dwelling people in the DRC, charcoal[16] is the only source of energy. For these communities, access to energy is what they need. It is iconoclastic to tell people not to cut forests for charcoal if this is the only source of energy. What is needed for people in these circumstances is first to access to more energy, not less. If that energy comes from renewable sources, that is second in their priorities and will be even more attractive. That for these communities what is needed is not a reduced consumption of energy but more energy isn't only attractive; it offers opportunities for new businesses and newer economy forms, which consist of less CO_2-emitting industries. It is also expectable that less industrialized countries may well be forced to emit CO_2 in some large quantities to reach the right levels of sustainable development before embarking on green economy. If there is any right to emit CO_2, it should be reserved to the world's poorest countries in their efforts to breach the chains of abhorrent poverty.

Greening Human Activities May Need a Little More CO_2 Emissions by Local Poor Communities

Accessing more energy is just one example; there are many more areas where what local poor communities need is not frugality but increments in their consumption to become stronger to fight climate change. Frugality (Adorno's simplicity) becomes holy only when the poorest people in countries that plant new forests, reforest degraded forests and protect forests are brought above the bare life natural necessities. These primary life necessities include food, clean drinking water, shelter, energy and good health.

That, in some circumstances, what is needed is a little bit more than pressing equal demands to cut consumption on everyone and everywhere necessitates a wider discussion on the optimal ways to address layered human needs. Human needs are not equal, and demands to address the global climate crisis should not be evenly distributed across the world. In a world where an American newly born baby costs 2 times more than a Swedish, 3 times more than an Italian, 13 times more than a Brazilian, 35 times more than an Indian, 140 times more than Bangladeshi and 280 times more than a Chadian baby,[17] there is no point asking everyone to make the same efforts to stop the climate from deteriorating. Inequalities in the usage of resources across the world are wider than it first appears when it is discussed in the world's powerful nations' narrative. Addressing climate change goes beyond the climate physics and poses questions of rights, morality and the question of the meanings of life itself (Chapter 6). For some world's parts, the grandiloquence of ethics, meanings of life, morality and rights does not mean anything. For Congolese, the obvious question is simple: how to survive before even asking questions of decency of life. This is why allowing poor Congolese communities to have a little more than asking them to give up something they barely have is not shallow ecology[18] as it was argued against the argument that local communities should be allowed to access natural resources in areas where they inhabit.[19] If justice and compassion exist in deep ecology, they should not demand more sacrifices from poor communities to pay the price of the opulence of the wealthiest nations. In circumstances where there is no malls and super stores with everything one needs and where the purchasing power of citizens is symbolic, shallow or deep ecologies are farfetched. This is not to argue that rural poor communities in Central Africa don't care about their environments; if anything serious can be expected of what ecology is, it should be to help these communities be able to use their natural resources in ways that will maintain the availability of these resources over a long-term perspective. This is sustainable development. The main question that is often asked in the DRC is how to stay alive and how to ensure that one's biological offspring will continue to live on the same resources in the future. For this to happen, there is no need for ecology to be shallow or deep. Ecology as a science doesn't speak the language of shallow or deep ecologies; what it does is primarily to try to disentangle and bring about the knowledge on relationships between living organisms and links living organisms have with their physical environments. It is comprehensible that writing from different perspectives, the same story is told in different tones. Any story always bears the imprint of where different storytellers stand; one writes only from where one stands! For those with no poverty experience, it is easy to transform ecology into a religion where everything needs to be kept away from human reach. Religious ecology is fine as long as it doesn't prevent other voices being heard. The voices of poor Congolese don't need sophisticated arguments; staying alive is what their first priority is. Their second priority is the will to give meanings to their own lives and strive to achieve the goals they have assigned to themselves. This is why frugality makes sense only when people have energy, food, good health and shelter. In current DRC conditions, frugality means rising up life standards for the poorest in the communities, which doubtlessly, goes along with increasing some CO_2 emissions for a limited time. This isn't to ask Central Africa to escalate its CO_2 emissions. Rising up living standards *also* means investing in renewable energy, producing short-traveled distance agricultural products, etc. The efforts to construct grids of renewable energy and produce and consume locally, in turn, need significant research and innovation efforts. For Central Africa, at this point, decarbonizing economic growth is luxury to be prepared for.

Addressing climate change poses questions of rights and morality and also questions the meanings of life itself. The question of the right to development was addressed above (Chapter 2). The morality of asking similar efforts from people who live under vastly differentiated economic conditions to fix the nightmares humanity created with the world climate was discussed in the last section of this chapter. Chapter 6 discusses the question of meaning of life. It takes on from the last point of this chapter where ecology is somewhat questioned. Chapter 6 aims at discussing perspectives that brought this book to the conclusion that ecology, as a science, should be neither shallow nor deep; it should become neither a religion of opulence nor a religion of unnecessary material asceticism, which imposes similar restrictions on all humans, irrespective of their specific conditions. In having that discussion, it is hoped that one would find a way to bridge the gap between ecological material asceticism and an un-thoughtful CO_2 emissions liberalism.

Notes

1 Harvey H, Orvis R, Rissman J (2018) Designing Climate Solutions: A Policy Guide for Low-carbon Energy. Islands Press, p. 376.
2 Blokhin A, Smith A, Pereze Y (2023) The 5 Countries That Produce the Most Carbon Dioxide (CO_2). https://www.investopedia.com/articles/investing/092915/5-countries-produce-most-carbon-dioxide-co2.asp
3 Fitzgerald J (2020) Greenovation: Urban Leadership on Climate Change. Oxford University Press, p. 256.
4 Swingland IR (2013) Capturing Carbon and Conserving Biodiversity: The Market Approach. Earthscan, p. 368.
5 Kill J, Ozinga S, Pavett S, Wainwright R (2010) Trading Carbon: How It Works and Why It Is Controversial. FERN Publication, p. 118.
6 Manning WJ (2020) Trees and Global Warming: The Role of Forests in Cooling and Warming the Atmosphere. Cambridge University Press, p. 338.
7 Wu X, Liu G, Bao Q (2023) Impact of economic growth on the changes in forest resources in Inner Mongolia of China. Frontiers in Environmental Science 11: 1241703.
8 Sandel MJ (2012) What Money Can't Buy: The Moral Limits of Markets. Farrar, Strauss and Giroux, p. 244.
9 Sandel MJ (2005) Public Philosophy: Essays on Morality in Politics. Harvard University Press, p. 292.
10 Grinin L, Korotayev A, Tausch A (2016) Economic Cycles, Crises, and the Global Periphery. Springer, p. 283.
11 Reuters (2022) U.S. asks Congo to pull some oil blocks from auction to protect forests. https://www.reuters.com/business/sustainable-business/us-asks-congo-pull-some-oil-blocks-auction-protect-forests-2022-10-04/
12 White LJT, Bazaiba EM, Ndongo JD, Matondo R, Soudan-Nonault A, Ngomanda A, Averti IS, Ewango CEN, Sonké B, Lewis SL (2021) Congo Basin rainforest — Invest US$150 million in science. Nature 598: 411–414.
13 Roychowdhury A (2018) Clunkered: Combating Dumping of Used Vehicles. Roadmap for Africa and South Asia. Centre for Science and Environment, p. 132.

14 Clapp J (1994) Africa, NGOs, and the international toxic waste trade. The Journal of Environment & Development 3 (2): 17–46.
15 Adorno T (1951) Minima Moralia: Reflections from Damaged Life. Create Space Independent Publishing Platform, p. 275.
16 Schure J, Ingram V, Akalakou-Mayimba C (2011) Bois énergie en RDC : Analyse de la filière des villes de Kinshasa et de Kisangani, Projet Makala, CIRAD, CIFOR, p. 88.
17 De Rivero O (2003) Le mythe du développement (Traduit de l'Espagnol par Robitaille R). Enjeux Planète, p. 241.
18 Kopnina H, Gray J, Lynn W, Heister A Raghav Srivastava R (2022) Uniting Ecocentric and Animal Ethics: Combining Non-Anthropocentric Approaches in Conservation and the Care of Domestic Animals. Ethics, Policy & Environment, https://doi.org/10.1080/21550085.2022.2127295.
19 Inogwabini BI (2020) Reconciling Human Needs and Conserving Biodiversity: Large Landscapes as a New Conservation Paradigm. Springer, pp. 355–382.

6 Biodiversity Erosion, Climate Change and Life's Purpose

Introduction

Since the early 1950s, humanity witnesses massive and unprecedented losses of biodiversity world-widely.[1] This biodiversity loss across the world in historic proportions is biodiversity erosion means. Rapid biodiversity disappearance is also part of the ecological crisis, which some iconic outspoken figures (e.g., Pope Francis[2] and Murray Bookshin[3]) express. The ecological crisis is attributable, in many respects, to the addiction to the products of ever-changing modernity.[4] To respond to this ecological crisis, several thought currents emerged to practically save what can still be saved. The myriad ecological thoughts on the urgency to safeguard biodiversity flourish in economics, industry and political theory; they flourish even in philosophical and religious circles. All these thinking lines, except a minority of skeptics, join in the idea that safeguarding biodiversity is to preserve the earth. Preserving biodiversity is a priority even if doing so will not take us back to the yesteryear's biodiversity states. Minimally, biodiversity should be preserved in its current good and functioning states, which can succeed only if and when nature is purged of cultural and technological flaws that make humans exert excessive pressure on the rest of the material and biological worlds. While the earth could have existed, without life as we know it or with other forms of life, the biodiversity conservation question is essentially a matter of preserving life in its current forms on the earth; it is essentially a question of relationships.[5] Protecting the earth becomes a priority to preserve humans-earth relationship through the living beings and functions that make human's life possible.

This chapter discusses the meaning of life using the lenses of the unprecedented and on-going global biodiversity crisis. The question that one can ask is why do we have to have entire chapter discussing the global biodiversity crisis in a book about climate change? Shortly answered, even though several factors cause the current biodiversity erosion, the effects of global climate change are the fundamental challenge that life itself faces. Climate change discussions are in reality debates about the diversity of life's future when

DOI: 10.4324/9781003493754-7

climate conditions dramatically change. Climate change is one major issue humanity faces. Therefore, asking the question of the meaning of life under current climatic conditions comes to reading the life's meaning through the optics of the global biodiversity. Before getting to the above question about the meaning of life itself, this chapter discusses the biodiversity concept and implications of any fundamental aspect of what biodiversity really is. This lays the ground before the critical question of the meaning of life is appraised.

The Biodiversity Concept and Its Implications

The biodiversity alarming erosion demands urgent action; what beliefs should guide that action needs an open-minded reflection, which currently abounds. Biological and ecological narratives revisited the biodiversity concept's definition. Biodiversity is no longer the single idea of species numbers as formerly. Biodiversity encompasses the species, the living spaces and the functions that both physical and biotic environments play to maintain life. Various life forms are cascades of inter-linked biological interdependences. Importantly, biodiversity also includes functional relationships that any element of the living whole maintains, reciprocally, with other living organisms and its environment.[6,7] In short, biodiversity is extended to include ecological biological, physical and social environmental functions that make life and how life sustains itself. Thus extended, the biodiversity concept questions not only the intrinsic value of species and the ecosystems in which the species evolve but also life itself in its various forms.[8]

As suggested above, the biodiversity issue goes beyond its biological foundations; it also has to be understood anthropologically and philosophically. Biodiversity contributes to the wealth of nations; it should also be discussed economically and socially. However, before economic and social questions are discussed, questions such as why biodiversity should be conserved, what type of biodiversity is worth conserving and, even, why these questions should be asked are also important.[9] Formulating questions such as why to preserve biodiversity and what type of biodiversity should be protected came from natural sciences. Biologists also extended the biodiversity concept to cover an area wider than just numbers of species. This is both understandable and unfortunate. It's understandable that biologists ask questions about biodiversity and extend its radius of comprehension because biologists daily work on this subject; they construct their careers gaining much deepened knowledge of how life works. It is unfortunate thinking biologists to be uniquely positioned to respond to these complex questions because they cannot find adequate formulations and answers within the closed and unique framework of natural sciences. Questions such as does life have any meaning or why one should be asked to preserve biodiversity need a broader reflection framework. This reflection framework goes beyond the ordinary channels of natural sciences.

An Open Biodiversity Reflection Framework

Thinking about biodiversity within a more extended philosophical and ethical reflection framework and epistemologically outside the usual biology's intellectual means is more necessary than usual at the moment when the question of why humanity should conserve biodiversity becomes a question of life. As indicated above, biodiversity conceptually includes physical and biotic and functional responses abiotic environments give to biotic solicitations. The inter-section of the biotic and the abiotic pulses plays a critical role in maintaining life; the relationships between the abiotic environment and biotic constituents of the same environment are integrated in a whole that maintains life.[6,7] There is reciprocity wherein living organisms solicit the physical environments wherein the organisms live and vice-versa. Responding to questions biodiversity raises should be done as a complementary cross-fertilization process between natural sciences, social sciences and humanities. The concept of biodiversity, after being extended to include the ecological functions of the physical, social and biological environment that make life maintain and support itself, raises questions not only of the intrinsic value of species[8] and the ecosystems in which they evolve but also life itself in its various forms. These various life forms are not reducible to flows of biological interdependent energies; they bear meanings above their current utility. Looking at how plants directly capture and use solar energy and CO_2 to create nutrients (Chapter 5), one wonders why this is so. The question 'why this is so' transgresses the natural sciences' epistemology, whose questions come in the form of how things work or how things are the way they are. Additionally, the question why humans should conserve biodiversity when biodiversity is the largest resource some human communities depend on to live is even a more pressing inquiry. This is understandable when some communities cannot cope with life otherwise but through extracting food directly from biological resources. The question whether people should extract food from biological resources is a social and an economic one before becoming a question of comparing the worth of human life with that of the species humans eat to live. Talking about biodiversity touches things that people use for their livelihoods and the material substrates people have been using over millennia to ground imaginaries and whole cultures. Talking about biodiversity, therefore, steeply descends to scientific multi-disciplinary. Natural sciences may offer the insights they gather on how biodiversity works, yet anthropology, cultural studies, economics, philosophy and sociology must come in the game to ensure that all perspectives are present to create a near total picture of the reality of biodiversity. This is why to reflect ton biodiversity requires a framework that is entirely open to influences from all areas of enquiry; it does even need to bring in artists to present their view of things they include in the concept biodiversity.

Biologists and Biodiversity

A large part of scientists answer the question of why we should conserve biodiversity by following biophilism. Edward O. Wilson[10] defines Biophilism as the love for life. The most plausible argument why humans should conserve and pay attention to biodiversity is because humans deeply love life itself. The attachment to life that humans feel irresistibly justifies the demand to conserve biodiversity. That the desire to preserve biodiversity is the love for and attachment to life itself opens the door to several questions. For example, if one wants to protect biodiversity out of love for life, why would humans love life so much? Feeling the duty to preserve biodiversity through love and attachment to life itself raises questions about the meaning and life's purpose. In natural sciences, Wilson is not alone to use such a language. James Lovelock also deliberately uses a similar language in his many writings on Gaia. Both Wilson and Lovelock's storylines form narratives that suggest a life with both intrinsic and extrinsic ends inscribed in a direction that projects life beyond itself. This is, in a sense, that life is independent of human needs, wishes, and activities.

When Lovelock[11] writes that humans should love and respect the earth with the same intensity as the love they give to their families and tribes, he gives the earth life attributes that natural sciences would not logically offer to natural physical structures such soil, rocks and water. Morin[12] define life, *tout court*, from the same repertoire when he says that life is feeling, affectivity and love. With such a humanizing language, it is not surprising that Lovelock[13] recognized that theorizing Gaia, the idea that the earth is capable of regulating its climate and chemistry to make itself comfortable for its inhabitants, reached at no scientific consensus initially. This is understandable since even today, matching life to a force, even a geological one, is vitalism. Vitalism is hardly acceptable in mainstream natural sciences.

Wilson and Lovelock insisted that the scientific community should think about life and biodiversity within a reflection framework broader than the natural sciences. Wilson and Lovelock were the intellectual breeds that can venture outside and above their narrow career paths to give rise to a rich variety of biodiversity conservation discourses. The renewed biodiversity conservation discourses can all be grouped into two major classes: First, the natural sciences' biodiversity conservation discourse; second, (different) nature philosophies. Nature philosophies bring together a range of thinkers from several horizons, ranging from ecological religiosity tendencies to academic philosophy of nature. This recalls the opposition between John Muir's views and Gifford Pinchot's ideas. Muir and Pinchot were, each in their own right, the pioneers of biodiversity conservation in its modern form. Pinchot supported a prudent use of nature, as opposed to the misuse of nature.[14] Muir argued for a transcendental view of biodiversity. This transcendental view of biodiversity meant that humans were compelled to use biodiversity only for cases that are too restricted.[14] For John Muir, humans could use biodiversity only in cases where higher motives to justify themselves to do so exist.

The natural sciences' biodiversity conservation discourse is traditionally rooted in quantitative methods for which the biodiversity erosion is explained by trying to resolve quantifiable questions such as the number of species, species space occupancy and species biological and geographical connectedness. The natural scientific discourse calculates the natural ecological niche's yields and the total earth carrying capacity. Quantitative discourse explaining doesn't rhyme with causal explanation. Pinchot's views were part of this tradition. Pinchot suggested prudently using nature. This view eclipses the idea that nature would be fully or extensively described and quantified so that levels of prudential off-takes can be carved from nature and be used to create better human living conditions. Pinchot opposed the misuse and abuse of nature, which is a historical precedent of concepts of rational natural resources use. Today, sustainable usage replaced the rational natural resources use. Margins between rational and sustainable usages seem so thin that the dispute about which jargon is appropriate is more or less but quarrels of chapels. Purposely, the adjective transcendental is used to express views Muir had for biodiversity. Muir viewed biodiversity with the fervor some ecological spiritualists have today. It is misleading to affirm that Lovelock thought that biodiversity had transcendental meaning more than just regulating itself, but Wilson seems to think things in their long-lost distant origin; Wilson was brought up in a Christian family. Muir was much closer to the ideas of something going beyond biophilism, which supports ecologism or the nature religion.

Biologism versus Vitalism

Muir and Pinchot's followers agreed on the importance to preserve biodiversity. Nowadays, nearly everyone agrees on this too. However, they fundamentally disagree on why biodiversity should be preserved. First, for natural scientists, the reason to conserve biodiversity is biophilia *qua biophilium*. However, philosophers of science, such as Desjardins,[15] add to the love of life for its own sake (biophilia *qua biophilium*) other reasons, including the beauty of life, biodiversity, economic utility, ecological culturalism and intergenerational generosity. But the foundations remain widely the love of life itself. Muir's followers (nature religious) view biophilism as a stepping stone to transcendental ends. These two discourses run in parallel and occupy the academic and the public spaces.

To reconcile these two discourses is historically a very daunting exercise. Thomas Aquinas attempted to synthesize the views that (1) nature was created and ordered in such a way that it should accomplish a teleological project and that (2) nature was uncreated and only responded to the laws of physics. During the time of Thomas Aquinas, physics was a philosophy of nature. Unfortunately, his work was satisfactory until the rise of probabilistic statistics shook the entireness of this monumental edifice. Introducing probabilities in natural sciences shook the foundations of classical mechanics, which was believed to give meaning to the causality of observable macroscopic phenomena and thomistic synthesis that relied on it. Probabilities brought new scientific approaches generally and the life sciences particularly. Instead of classical causality relations, statistical sciences introduced the notion of chance. Statistical chance postulates the non-existence of magic causal hands. Statistical approaches admit of no magic causal hand, which would arrange things and organize life for purposes other than what people see and live by.[16] With statistical approaches, the world is driven without a big plan behind it. The positive evolution of scientific thought in the 20th century generally led to mind-openness; however, causality and nature teleological direction also remain anchored in human thought and unfolds in parallel with the statistical explanations of things. For statistical sciences, chance (probabilistic acceptance) alone explains how things work or they are made to work the way humans perceive them. Perceived physical facts (events) occur by chance; their respective chances of appearances are linked to probabilities without an extra natural cause.

Explaining things by chance or teleology runs in parallel, and parallelism means that a common discourse on the ends each group pursues seems out of reach. Exploring a possibility to reconcile the two ways to understand things seems incommensurable. For biologists, thinking of a purpose other than biophilia is guilty of Bergsonian Vitalism.[17] Vitalism has no place in the search for scientific truth because it is methodologically imbued with intuitions[18] to which biology gives little credential. Scientists would agree that intuitions offer access to some knowledge forms. However, the knowledge forms accessed through intuitions are difficult to assess thoroughly even though Bergson argued that immediate intuition and discursive thought do one thing. Vitalism is often unwelcome to biologists because it seems to include spiritualist tendencies.[3] Natural spiritualist tendencies are rooted in Baruch Spinoza's pan-vitalism in which everything that exists has a signature of life behind it. Spinoza[19] wrote (proposition 21) that God is an immanent and not a transitive cause of all things. A more globalizing formula is the one [...] which makes space and time essential determinations of the first Being itself, but also looks at the things that depend on this being simple accidents inherent in it, and not as substances [...].[20] Élan Vital is a self-organization principle and the ordering force in nature. This definition explains why Élan Vital is straightforwardly repelled by biologists.[3] Biologists, chemists, mathematicians and physicists are attached to the idea that life does not need anything like teleological direction to explain itself. For these natural scientists, life is essentially a fortuitous fact in the natural evolutionary history.

On the other hand, biologism limits the conditions to acquire extended knowledge on life and it remains too reductionist.[21] Biologism traces everything back to organic and inorganic substrates of life and its evolution. For critiques, an overflow of biologism becomes in itself an ideology.[20] This is a more acute criticism now that biodiversity encompasses not only species but also functions.[6,7] The discourse some natural scientists sometimes make has teleological overtones. Two examples of such scientists with a confusing language, if not carefully read through, are Wilson and Lovelock. The biologistic discourse on biodiversity ignores the teleological effluvia inherent in biologism itself. Several modern biologists who are converted to the cause of biodiversity and of life itself, reintroduced final causality using secular terms that make deciphering causality very tough. Nagel[22] was right that from the moment functional explanations are introduced into the biological approach, the difference between scientific explanation and recourse to teleology [as a possible alternative] explanation comes from selective attention rather than the content of the statements being made. Sure, there is a dichotomy between scientific culture and the culture of the humanities, a differentiation that Morin[7] clarified. Humanities offer a general culture that allows reflection on scientific work. This reflection includes works of ancients on human conditions and the destiny of human beings. Scientific culture is an ethos of specialization where knowledge is compartmentalized and where the ability to reflection on knowledge itself is [nearly] absent. Morin's remark on the near absence of reflective thought in the natural sciences is strongly and rightly questionable, but his presentation of this dualism isn't; it produces an adequate analysis with regard to the biological sciences and the question of what life itself is. The analysis seems adequate because there are very few people working on biodiversity who think beyond species, functions, habitats and inter-relationships between these entities.

The mutually *de facto* exclusion of biological sciences and social sciences makes the dialogue difficult in the practice of biodiversity conservation. Parallel discourses sometimes confuse the discussion of why biodiversity should be conserved. Evident confusions over narrative often negatively impinge on the day-to-day conservation job. Temptations to find solutions to the fundamental question of why to conserve biodiversity can be seen in the work of biologists[23] in which intriguing hypotheses such as the argument that even the emergence of religion can also be mapped into genetic dimensions and evolutionary change. Can one imagine? In linking genetics and religion, Wilson was arguing that the teleological worldview was part of the genetic human evolution. The biodiversity value was also taken up by a certain number of professional philosophers and other humanists. One relevant example to cite is Ronald Dworking[24] who introduced differential levels of the life sacredness, providing an argument for classifying species from the most sacred to the less sacred ones. The more sacred species deserve to be preserved while lesser sacred species deserve less conservation efforts.

The above overview shows that the natural science discourse and the humanities' narrative are still far from reaching a common ground on questions like what biodiversity is, why biodiversity is important, and why it should be preserved. All parties discussing biodiversity conservation call for biodiversity to be preserved. Nonetheless, when it comes to why to do so and what biodiversity deserves such special attention, answers widely diverge. Interestingly, the divergence remains wide even when all parties resort to the analogous semantic repertoire. This is witness to the need to acquire a common understanding of what shared concepts, idioms and metaphors truly mean. In other words, the main question about the meanings, values and how to sustainably use renewable natural resources is to respond to the call for an open perspective.[7] This open perspective on biodiversity is more complex but would provide solid foundations to an enduring and stable relationship between biology and humanities and social sciences. Yet, this relationship is to be at the same time complementary, competing and, even at many points, antagonistic.

Outlining an open, complex perspective that would establish an indissoluble, permanent, complementary, competing and antagonistic relationship between conservation biology and the promises of human sciences needs to be not only a multidisciplinary effort but also an effort wherein cultural perspectives and scales of valuing things across the world are factored. It is necessary to understand the natural science facets and the human science aspects of biodiversity conservation. Conducting such an inquiry should begin by clarifying key concepts these parallel discourses most commonly use. It is useful to recall that concepts such as biodiversity, environment and nature don't have the same semantic subtleties and economies in natural sciences and humanities and philosophy. For example, Aldo Léopold takes nature as a complex but highly organized and stable whole that is crossed by energy flows[25]; Immanuel Kant defines nature as a set of all the objects of experience [constituting] a system according to transcendental laws […].[26] Furthermore, biodiversity, environment and nature don't even factually include views of some world's parts; they are essentially Western world's constructs. Ironically, it is in these ignored parts of the world that the Western-conceived conservation biology and climate change are tested; the ignored parts of the world are the ones from which efforts to curb the current biodiversity and climate change crises are being asked to be made.

Different views lead to divergent perceptions and influence the resulting analysis. To avoid the reduction's trap dictionary interpretation imposes on the words, clarifying key biodiversity concepts has to be a critique that is not only semantic but also social. This critique should rely on examples the practice of the conservation of biodiversity inspires. Quoting Leopold and Kant back-to-back, as above, points to the other methodological direction an inquiry to bring natural sciences and humanities closer should take. Inquiring about the meaning of biodiversity in the context of a changing climate should be done in a back-to-back reading of the current fundamental texts. These fundamental texts should be chosen with regard to their leaning toward naturalism or teleological approaches to biodiversity conservation. The first fundamental text of naturalism would be along the lines of 'Conservation Biology: The Science of Scarcity and Diversity' that Soulé edited.[27] Many conservation biology scholars considered Soulé's work to lay the foundations of this new biology branch focused on nature and its diversity. Soulé's text should be studied back-to-back, for example, with the 'Praised are you',[28] which foundational text addresses biodiversity conservation in some parts. These parts should be critically examined using not only novel scientific knowledge but the past foundations on which this knowledge is built upon. The force of the natural disasters over the past decade made Pope Francis' 'Praised are you' too famous since 2015 when it was first published. Indeed, since then, people across the world have been mourning the ones they dearly cherished because of natural climate change catastrophes. Felt threat may heighten human conscience (Chapter 1).

Reading back-to-back Soulé's and Francis' texts, for example, and reading them in the light of the clarifications and positions that each took would further clarify the texts and enable comprehensible course of action. Reading back-to-back is a critical effort to understand two visions of biodiversity conservation, one scientific and the other teleological. The back-to-back comparison leads to re-reading texts cross-referentially. Using positions other thinkers took, a thinking gradient emerges, ranging from strong naturalism to strong biodiversity conservation teleology. Foreseeably, the back-to-back and the cross-referenced reading of the two primary authors makes it possible to identify the points of rapprochement between the two parallel biodiversity visions. Interpreting these rapprochement's points open-mindedly results in a more enlarged comprehension of and more palatable decisions about what biodiversity really is and why we should conserve it. The open perspective should not unnecessarily sacrifice the complexity of each perspective and the complexity of the new whole as it emerges while forming a permanent dialogue between parallel ways to understand the world. The reality is that biology complements, competes with and antagonizes human sciences, and vice-versa.[29]

Where Is Central Africa's Voice in This Discussion?

Biodiversity conservation is a question of what type of the world humans want to live in for the next decades. A literature review of what is cited thus far clearly indicates whose voice is present where the world's future is discussed and defined. After summarily discussing biodiversity conservation, it is clear that voices from some world's parts barely exist and much less heard. Stating that the voices of a world's large part are absent in this world's future-shaping debate doesn't mean that there are no people from Africa,

Asia and Latin America who are involved in ecological research. Numbers of people from these areas attend many meetings that are mostly international biodiversity and climate change masses. Diplomacy being still diplomacy, delegations from ignored world's parts attend these masses (Conference of Parties [COPs]) as token-representation (Chapter 2). For sure, the names of researchers from ignored parts of the world appear on traditional scientific peer-reviewed papers or chapters of massively important books. So, claiming that voices of some world's parts are ignored in meetings that shape the world's future doesn't mean that if the views of voiceless communities were brought to where they can be listened to, it would necessarily convey different messages. It might just be that the voiceless would come to the same conclusions and adopt similar strategies. But the absence of the voices of the voiceless means a twofold reality. First, the current science practice traces its origins to the Western thinking system. This is why one sees someone writing from Central Africa and about Central Africa extensively quoting Leopold, Kant and Aquinas, as I have done throughout this book. I am not implying that quoting Leopold and others should be a motive of shame. There is no reason for this to be the case because the knowledge that has been brought from the ancient Greeks down to Soulé is the knowledge that humanity collectively created over the course history.[30] As such, it contains the knowledge that came from diverse world's parts. There is no question that knowledge has been magnified and stored in the Western world in its current usable format. However, the bottom-line is that the knowledge humanity currently possesses is from human collective efforts; there is no guilt to think the world from Central Africa with Leopold, Kant and Aquinas. Second, social and human sciences are about the primary location of power struggles; the ideas of the politically most powerful are the ones that are carried through, whereas the voices of the less powerful are silenced. Political power influences the writing and the dispatching of knowledge across the world. This is true since the 1960s, when formerly colonized countries began searching for the meaning of the history and the stories of their countries in narratives written by colonizing powers. Morris,[31] Spivak[32] and Gilroy[33] reflect on how political power imbalances silence the most powerless. The powerful group's knowledge is the knowledge par excellence while the knowledge of the powerless is robbed; otherwise, it is sweetened to suit the views of the most powerful. If the knowledge of the powerless displeases the politically powerful, it is either ignored or the power-to-be brutally erase it. This statement has no uniqueness; what is new in it is that this reality prevails even during biodiversity erosion and climate change grand crises. When both crises threaten life altogether, the surprise is to find out that power struggle is still the same when one would be expecting life-threatening crises to bring people closer as never before. Unfortunately, political narratives on global crisis are limited within the rooms of climate COPs (Chapter 1). In the global biodiversity conservation and climate change debates, Central Africa's voice is rather dim to hear in discussions. The reason for this is that there are limited scientific capacities to handle the issues from Central Africa (Chapters 3 and 7). Political power imbalances across the continents are another reason. Power doesn't include intellectual power, along with financial, positional and the (often) the brute might Biodiversity erosion and climate crises manifest the fact that industrialized countries exert all the powers they have to get their agendas accepted by all. There are many problems with this way of envisaging things, but the chief problem is that industrialized countries equate the might with the right.[34]

What Is It that Would Look like a Central Africa's Voice in This Discussion?

The bare truth is that the voice from Central Africa is very little included in the mainstream biodiversity conservation and climate change debates. It is striking that even loadable African efforts[35] do not include samples from Central Africa. Far from signaling the absence of ideas in Central Africa, this is an epitome of a general trend, which is that most debates about biodiversity (forest, wildlife species, and freshwaters of the Congo Basin) and climate change are conducted on the backs of people who are not fully informed and whose contributions are not sought. Worst of all, powerful people don't often even think of contributions from the powerless.

If Chimakonam and his colleagues' work[35] reverberates the minimal common denominator of pan-African biodiversity conservation philosophy, therein are conclusions mimicking Central Africa's reality. Primarily, the conclusion that 'of theoretical approaches to moral status [of biodiversity], the idea that there is a hierarchy, and that the hierarchy is a function of ability to consume seems [...] the most promising way to philosophically ground animal rights [...]'.[36] This brings us back to the discussion on shallow ecology and deep ecology and the conclusion that for most of Central African citizens who rely on natural biological resources to live, asking them to give up something to protect biodiversity equates to asking them to starve to death[37] (Chapter 5). For Central Africa's communities, the basics of any conservation have meaning but that meaning is through hierarchy of values across a wide range of life forms. The hierarchizing process ranges from humans, sentient animals and plants. Hierarchizing biodiversity has the merit to avoid randomization of which life to sacrifice which would be the logical outcome of species egalitarianism that appeals to unqualified attribution of intrinsic properties such as being sentient, being able to flourish, being a subject of life.[34]

Hierarchizing different biodiversity parts doesn't reject the idea of love for life itself (biophilia). If anything can be said about citizens of Central Africa, risking repeating a cliché, it is that they just love life. Central African citizens love life abundantly and live it intensively and intensely.[38] But they don't do this only *qua biophilium*; there are other reasons to love life. One reason is the liaison with the world of ancestors in a continuum of life where the idea of past, present and future generations is granted in its own right (Chapter 4). No wonder, the possibility of [appealing to] moral duties to future generations is a meaningful way to develop a plausible account of the moral obligation to preserve species and biodiversity.[39] The other reason is that Central Africa would preserve its unique biodiversity if biodiversity holds a potential to generate important and sustainable financial revenues in the future.[39]

Citizens of Central Africa are half-way between purely biologistic life views and a deeply teleological world thinking ways. Let's note that there is a multitude of teleology. One version of teleology is the strongest version is finalism (external telos), whereby providential plans accompany everything that exists. In this version, teleology posits the existence of supernatural design intelligence.[40] Finalism is the watch-maker theology; when one finds a watch anywhere, one ought to think about the watch-maker whose work one sees in the actual watch.[41] The finalism is the teleology theological traditions developed and sustain. Theological traditions being but one of many ways to look at the world, there are other ways to formulate teleology. Teleology is the science of goals, purposes and meaning that people attribute to existing things. It does not always trace things back to a supernatural intelligence. This is so because there are goals and purposes that don't necessarily have to do with a grand design. There are other forms of telos. For example, the telos of eating food is to keep the living organisms alive. Biophilia constitutes explains the biologistic world view. Biophilia is an obstinate love that people have for nature because of its beauty and perfection. With this assertion, it is difficult to avoid seeing in biophilia reminiscences of teleological world's views in biologistic life projections.

Stating that citizens of Central Africa half-way between purely biologistic life views and a deeply teleological world thinking ways comes down to several points. The first point is that the citizens of Central Africa love the nature around them. This nature is made of forests, savannahs, rivers, lakes and other natural scenes such as mountains one finds across the region. However, this statement should be immediately re-qualified to indicate that the love is not that one indolently observes the world around. Forthrightly, the love citizens of Central Africa have for the nature around them isn't and shouldn't be confused with a mystical relationship with nature wherein people are immersed in a nature without taking any active action to make natural things work for their desires and life. The people of Central Africa are not poised in the thought that nature does its work for them and that they only have to accept whatever falls upon them. The love the citizens of Central Africa have for nature results from an objective relationship.[35] People of Central Africa depend on the ecosystems they dwell in to sustain themselves through their intellectual and physical works. This form of love for nature is, for example, described in the attachment Zande people have with their lands.[42] Without being singly economic in its value, land's current users love it and cannot alienate it; current land users cannot sell because it belonged to the three coexisting planes of existence. These planes were past, present and future; they include the dead who still exist, the living and the unborn would succeed us (Chapter 4). The love for nature is partly the love people have for their societies, including the deceased loved ones and those who will come later; the next generations are the continuation of the blood of people presently dwelling on the ancestors' land. This is teleology without a grandiose cosmic design, which has its meaning in that it helps people think of what they can do to achieve a meaningful life of their own. But land, as it has been found among the Dogon in West Africa, forms the first and foremost natural asset.[43] Therefore, it is through benefits people obtain from possessing the land, even if it is communally owned legally, that all the meaning-conferring mythologies were brought about. Bounties people obtained from the land may have created the telos that actual people attribute to it, which then turned the relationship into a strong filiation. This doesn't liquidate the fact that misfortunes, illnesses and deaths are also attributed to nature in the very Central African geographic and cultural contexts. Recourse to nature often explains witchcraft, sorcery, evil spirit, evil forces and evil powers.[44,45] Suggesting that bounties people obtained from nature preceded the status nature was given in the thoughts of people in Central Africa conveys the simple message that the imaginary Central Africa's people hold of nature is often crafted by the use of environmental conditions they live in. The material biological and non-biological environmental conditions shape the way people think and interact with the nature. This much is a truism; humans influence environments and environments influence the people they shelter. Other explanations, including strong finalism views, are also present in Central Africa. This is not surprising at all. These teleological explanations are evidence that Central Africa's people are capable of projecting themselves out in the space and time to abstractly think their lives and associated meanings beyond the physical environments.

Possible ecological and environmental imprints on human imaginaries in the Congo Basin aren't environmental determinism, which should be cautioned against. It is tricky to infer anthropological, cultural and social characteristics and traditions from biophysical environment.[46] The cultural and environmental cross-fertilizations are not straightforward linear relationships. Environmental determinism's theoretical framework emphasizes the importance of biophysical environments in determining social structures, economic progress and cultural development[39]. Jared Diamond[47] popularized the idea that geography determines success. Diamond argues that inequalities in the distribution of species and differences in habitat types largely explain inequalities in wealth and uneven potentials to develop.[48] Because cultures need material substrates to flourish, it seems plausible that interactions between humans and their immediate biophysical environments influence ways in which communities use physical materials to build material and symbolic cultures. Biophysical environment may partly explain some social behaviors, deeds and cohort of totems that communities hold dear and are integrated in collective identities. Ecological and environmental patterns can shed lights on some traits communities reveal as identities and ways they use to handle natural resources. Biophysical environment is not the only factor that influences human cultural development. Evolutionary biology suggests that human brain is formed to extract its activities above human body's biology and the physics; human brain often goes even beyond human conditions. Spirituality, laughter, joy and an understanding of the meaning of life come out of materialist brains.[49] Cultures are built through processes whereby sensible things are taken up by human brains and brought to an imaginary level where meanings are created. Life meaning is itself part of this image-generating process from what material life suggests to the brain. Wherever man and natural products are concerned, the idea intervenes. People know their habitats only through their own interpretation. Reality is always both cognitive and phenomenological[50]; realities are there for human communities to help humans make choices to cope with their environment. If anything, humans were not wired to behave robotically.[49]

Despite the cognitive love relationship people of Central Africa have for nature and life, they are not laid-back from using the biodiversity for legions of practices. The first and most apparent usage of biodiversity is the food people live on. People also use biological properties (functions) of biodiversity around them. Plants are used as medicines in multiple ways to heal people. Using the examples of food and medicines people use in Central Africa helps to understand why beauty and ecological habitat stability alone hold thinly as valuable justification to get people accept creating protected areas on the lands they rightly claim as the lands of their ancestors. Making citizens of Central Africa agree with the creation of protected areas on the lands they claim theirs by virtue of being inherited from their ancestors, natural beauty and ecological habitat stability arguments provide slander motivations. My own experience in Kahuzi-Biega National Park (Eastern DRC) in 1995–1996 speaks loudly about this. I led a community sensitization mission that encountered serious difficulties in convincing local communities along the road Itebero – Nzovu of the idea that the park was created for their own good. After three days with no progress, the team and I discussed the best ways to get the message that conservation was good, not only for the white people but also for everyone including themselves in their local villages and houses. Communities rightly demanded what their stake was in the presence of the park. The team agreed to tell communities that Kahuzi-Biega National Park was set aside outside of the reach of human activities to serve as the maternity space for the wildlife species. They would benefit from capturing legally authorized species outside of the park for their own food. These individual wildlife species in excess inside the park that would venture outside of the limits of the park should be captured by community members for community food only. 'Park as maternity space for wildlife species' is often used across Central Africa. The word 'only' was inserted in the message to suggest that wildlife trade was not part of things the communities were allowed to do. Communities were not as dumb as the technicalities of the biodiversity conservation messages suggested. At the end of each discussion, people always asked two important questions: (1) if what was said was true, why did law enforcers arrest people even when they captured wild animals outside of the park? (2) Why was selling meat to neighboring communities illegal, if capturing wildlife is only allowed for the local consumption? Local communities grasp the intellectual gymnastics this explanation takes to convince people plainly remain skeptical about its rationality.

Park as Maternity Spaces for Wildlife Species?

'Park as maternity space for wildlife species' says that notions brought from outside are often difficult to translate in local languages. This is not only for biodiversity conservation but also the difficulties to translate technical questions into local languages abound in climate change and sustainable development. Often, biodiversity conservation and climate change questions are translated in less precise idioms; for example, precise meanings sciences attribute to ecosystems, ecological relations, ecological valence, feedback loops, GHGs, albedo and residence time are finally lost. Biodiversity and climate change ideas are translated in long sentences that loosely say anything. The difficulty is not just about finding the right words; it is also about getting the people to understand what words mean and translate that meaning in how they perceive things. This doesn't mean that Central Africa's peoples are not aware of changes happening in their environment. People in Central Africa already feel dire consequences of regionally changing weather patterns. These language difficulties don't also imply that the people of Central Africa care less about ideas from outside. Governments and communities of Central Africa already absorb century-long assumptions over which international conservation organizations build the biodiversity conservation paradigms.[51] Finally, difficulties in languages don't mean that communities in Central Africa are apathetic and incapable of thinking of the plight they face in losing biodiversity and seeing climate globally changing. All 'park as maternity space for wildlife species' shows is that it is difficult to translate imported concepts. The translation process may leave imported concepts to bear different meanings, which can lead to different appreciations of field action's purpose. These difficulties tell us that global crises like biodiversity loss and climate change cannot be resolved by relying on one narrative. This is even more important when the narrative in question comes from just one source. Biodiversity erosion and climate change narratives emanate solely from Western societies. There is no objective problem to borrow the Western narrative, but there is also no moral obligation to reject the idea to contextualize solutions to account for local conditions. Contextualizing solutions passes through Central African scholars inventing new vocabulary to interrogate both biodiversity and climate crises. Languages are not stones idly lying unchanged over time. Languages are living entities[52]; they historically thrive by the capacities they have to borrow and absorb novel aphorisms and vocabularies from other languages.[53] Inventing new vocabulary to integrate the novelty of climate change in the speech and thoughts is not demanding too much for Central Africa. African languages extensively borrow technological innovation's vocabularies. Throughout times, the capacities of Central African languages to survive resided squarely in inventing new aphorisms, idioms and vocabulary.

What Central Africa needs to fight climate change re-emphasizes the idea that what is desired for materially poor communities is not less but bit more and better material conditions. This idea is plausible not only because it means to sustain human lives but also and primarily because it is the narrative people have of what is meaningful for them. Again, this is not to infer that people of Central Africa love nature just because of the bounties it offers them. Yet, the plausibility of having a little more and better life conditions through using natural resources does not also imply that the bounties nature can offer are inessential to lives people live. The need to materially sustain life itself and the love for nature and meaning it brings to the anthropological, cognitive, sociological and psychological well-being of people of Central Africa are part of the play with nature.

Assuredly, there is no question about whether Central Africa's communities know and appreciate the natural beauty and the goodness of the environment they thrive in. Certainly, people of Central Africa have the practical knowledge that enables them to interact with their environment. Beyond gathering and extracting food and other material goods, interactions people of Central Africa have with nature were described as part of the 'Force Vitale'[54] (vital force). It has been argued that Bantu philosophy was foremost Tempels own philosophy.[55] This cautions to uncritically apply vital force and its implications to citizens of Central Africa. Tempels was also said to be influenced by Bergson.[56] This means that 'vital force' simply transposes Bergson's Élan Vital to fit Central Africa's conditions. However, a significant difference seems to exist between Élan Vital and 'Force Vitale'. Élan Vital is a force distinct from the physical matter on which the former exercises its influence.[57] 'Force Vitale' is a force within the physical matter and vice-versa, making both a unique entity.[58]

In Central Africa, some plants and wildlife embody forces. In ceremonies, some plants and wildlife are addressed as if they hear, understand and can respond to humans in human language. Multiple requests of diverse nature are addressed to ancestors and other spirits that reside in plants and wildlife being addressed to. Despite bearing forces they symbolize, vital forces are external to plants and wildlife. Plants and wildlife only host forces. If Tempels meant that 'force is in matter and vice-versa and that they make the same and unique entity' also applied to humans, he missed a point. Subsequently, the implication that imaginarily 'humans and nature were the same thing because people bear force vitale' was flawed too. People in Central Africa don't see themselves as being singly part of nature; they are both part and outside of nature. Force Vitale is all the natural forces along with the forces the deceased ones have. These forces can augment the forces of living people. Using natural forces and the ancestors' powers is half-way between Vitalism and Finalism and the practical usages of natural things through applying the knowledge humans historically gathered. There is nothing new in stating this, but doing so is extremely valuable because the myth of primitive Africans being part of the nature is often used to justify why protected areas should be promoted in Central Africa. Unspoken theories go that because primitive Africans are part of the nature, forests should be left standing to prevent climate to escalate to changes that are irreversible. The language of the polished political correctness hardly uses words such as primitive Africans nowadays, but field activities are not so mute on this provided that one thinks beyond conventional discussions.

The last point above should not to be wrongly interpreted to mean that there should be no action to protect biodiversity in the Congo Basin. It doesn't promote actions against reforestation or preservation of standing forests in the Congo Basin. The whole point is that to preserve biodiversity in Central Africa, voices of the citizens of Central Africa should be heard. Realities people across Central Africa live in should be factored in schemes that are being crafted to be implemented in Central Africa; their ideas should count. Voices shouted out of Central Africa are those of materially needy people who want to have better material conditions. Realities they live in are miserable poverty; wishes they dearly hold on are to promote sustainable development. Trying to impose other ideas because of the global needs likens to the same story of the world imbalances in economic, military and political powers. The powerful can impose everything on the powerless: imposing ideas, technologies and even deciding on who should die and who should be kept alive.

Objections to the Claim that True Voices of Communities Aren't Truly Attended to

Objections to the statement that to preserve biodiversity in Central Africa and fight climate change, voices of local communities should be attended to, real needs of sustainable development should be factored in schemes to create conditions for the use of natural assets, are likely to stem from the argument that citizens of Central Africa are always present in discussions of globally important climate and biodiversity conservation questions. The argument would also be about the 'free, informed and prior consent' to obtain from communities before any action on global questions can be taken. Chapter 7 deals with these two objections; it is suffice to say that ecology is a branch of science. Ecology aims to acquire the knowledge to support the action by bringing together different thematic areas of scientific inquiry and knowledge. At its best ecology, whether deep, global or shallow, should be a conduit via which the multi-facets scientific definitions of biodiversity and nature go through before they acquire their respective full blown philosophical meanings. These meanings can be deontological, extrinsic, intrinsic, utilitarian, etc. Ecology, as a philosophical account and as a life practice, should be an account of how people in each context view the relationship they have with biodiversity and nature and how people perceive their place within nature. Communities give full meaning to the nature around them, including respect to nature, which is to put the boundaries between the permissible and the impermissible. Where people are tempted to go beyond the boundaries, education (scientific, philosophical, and religious traditions) can be used to sensitize them about nature's tolerance limits. This follows the traditions of having world-widely known scientists, Buddhists, philosophers, prophets, etc., who come and try to share their knowledge and intuitions with the rest of the world. This is how humanity globally transmitted the knowledge it cumulated over centuries from Archimedes to Penrose, from Socrates down to Rawls, from Buddha down to Dalai Lama and from Saint Francis of Assisi down to Pope Francis. Ecology, whether deep or shallow and whether global or partial, should not lead to ecologism, which becomes an ideology. Discussing ecologism as either an ideology or a religion goes beyond the scope of this chapter and will not be taken any further this point.

Deep (shallow), mild (strong) and global (partial) ecology alone offers a weak argument to prevent people from accessing natural resources to live on. Ecology cannot be a sufficient reason to prevent Central African people accessing natural resources to achieve material life sustainability when they starve. Ecology can, however, provide a sufficient reason to prevent people from going beyond the thresholds of the bearable. This is feasible only when both biodiversity and human needs are equally catered. Central Africa is the least industrialized and the least developed world regions. In circumstances like these, ecology should be a community scientific and philosophical education to define what is good for people and how to achieve the global public good without going beyond the limits of tolerance of nature. This education can be even a religious one (why not), if that is the right way to go for the communities in Central Africa. Ecology can also be anchored in African traditions, which are not all eco-friendly, contrary to most popular views. People, no matter where they are, are always tempted to go beyond natural boundaries nearly everywhere and whenever that is possible.

Notes

1 Kolbert E (2014) La 6ᵉ Extinction: une histoire contre nature (Traduit de l'Anglais par Michel Blanc). Nouveaux Horizons, p. 347.

2 Pape François (2015) Loué sois-tu (*Laudato Si*) : Lettre encyclique sur l'écologie. Edition Saint Augustin, p. 196.

3 Bookshin M (1990) The Philosophy of Social Ecology: Essays on Dialectical Naturalism. Black Rose Books Edition, p. 197.

4 Feyerabend P (1999) Conquest of Abundance: A Tale of Abstraction versus Richness of Being. The University of Chicago Press, p. 285.

5 Jonas H (2001) The Phenomenon of Life: Toward a Philosophical Biology. The Northwestern University Press, p. 303.

6 Morgan GJ (2009) The many dimensions of biodiversity. Studies in History and Philosophy of Biological and Biomedical Sciences 40: 235–238.

7 Bosworth A, Chaipraditkul N, Cheng MM, Gupta A, Junmookda K, Kadam P, Macer D, Millet C, Sangaroonthong J, Waller A (2011) Ethics and Biodiversity. Asia and Pacific Regional Bureau for Education. UNESCO, p. 117.

8 Naess A (1995) The deep ecological movement: Some philosophical aspects. In Sessions G (Editor) Deep Ecology for the Twenty-first Century. Shambhala: 64–84.

9 Minteer BA, Miller TR (2011) The new conservation debate: Ethical foundations, strategic trade-offs, and policy opportunities. Biological Conservation 144 (3): 945–947.

10 Wilson EO (2009). Biophilia. Harvard University Press, p. 168.

11 Lovelock J (1995) Gaia: A new look at life on earth. Oxford University Press, p. 148.

12 Morin E (2015) L'aventure de la méthode. Editions du Seuil, p. 159.

13 Lovelock J (1995) The ages of Gaia: A biography of our living earth. Oxford University Press, p. 255.

14 Callicott JB (1990) Whiter conservation ethics? Conservation Biology 4 (1): 15–20.

15 Desjardins JR (2013) Environmental ETHICS: An Introduction to Environmental Philosophy (5th Edition). Wadsworth, p. 179.

16 Kahneman D (2016) Système 1/Système 2 : les deux vitesses de la pensée (Traduit de l'Anglais par Clarinard R). Flammarion, p. 705.

17 Bergson H (2013) L'évolution créatrice. Presse Universitaire de France, p. 732.

18 Bergson H (2013) Essai sur les données immédiates de la conscience. Presse Universitaire de France, p. 340.

19 Spinoza B (2001) Ethics (Translated by White WH, revised by Stirling AH with an introduction by Garret D). Wordsworth, p. 276.

20 Kant E (1985) Critique de la raison pratique (traduit de l'Allemand par Ferry L et Wismann H). Editions Gallimard, p. 252.

21 Heinich N (2012) Naturalisme, antinaturalisme : non-dits et raisons du réductionnisme. Sociologies : Débats, Le naturalisme social. http://journals.openedition.org/sociologies/3814.

22 Nagel E (1979) The Structure of Science: Problems in the Logic of Explanation (2nd Edition). Hackett Publishing Company, p. 618.

23 Wilson EO (2004). On Human Nature (with a new preface). Harvard University Press, p. 260

24 Dworking R (1993) Life's Dominion: An Argument about Abortion and Euthanasia. Harper Collins Publisher, p. 272.

25 Leopold A (1949) A Sand County Almanac. Oxford University Press, p. 295.

26 Kant E (2000) Critique de la faculté de juger (Traduit de l'Allemand par Renaut A). Editions Flammarion, p. 535.

27 Soulé ME (Editor) (1986) Conservation Biology: The Science of Scarcity and Diversity. Sinauer Associates, p. 598.

28 Pape François (2015) Loué sois-tu (*Laudato Si*) : Lettre encyclique sur l'écologie, Edition Saint Augustin, p. 196.

29 Morin E (2023) Encore un moment...Textes personnels, politiques, sociologiques, philosophiques et littéraires. Denoel, p. 188.

30 Kodjo-Grandvaux S (2013) Philosophies Africains. Présence Africaine, p. 301.

31 Morris RC (Editor) (2010) Can the Subaltern Speak? Reflections on the History of an Idea. Columbia University Press, p. 318.

32 Spivak GC (1999) A Critique of Postcolonial Reason: Toward a History of Vanishing Present. Harvard University Press, p. 449.

33 Gilroy P (2006) Postcolonial Melancholia. Columbia University Press, p. 170.

34 Grayling AC (2023) Philosophy and Life: Exploring the Great Questions of How to Live. Penguin Random House, p. 426.

35 Chimakonam JO (Editor) (2018) African Philosophy and Environmental Conservation. Earthscan/Routledge, p. 234.

36 Metz M (2018) How to ground animal rights in African values: A constructive approach. In Chimakonam JO (Editor) (2018) African Philosophy and Environmental Conservation. Earthscan/Routledge, pp. 30–39.

37 Kopnina H, Gray J, Lynn W, Heister A Raghav Srivastava R (2022) Uniting ecocentric and animal ethics: Combining non-anthropocentric approaches in conservation and the care of domestic animals. Ethics, Policy & Environment 26: 265–286.

38 Mukashing RM (2019) L'Homme et la nature: Perspectives Africaines de l'Ecologie profonde. L'Harmattan, p. 308.

39 Behrens GK (2018) An African account of the moral obligation to preserve biodiversity. In Chimakonam JO (Editor) African Philosophy and environmental conservation. Earthscan/Routledge, pp. 42–57.

40 Bullock A, Tremble S (editors) (1997) The Fontana Dictionary of Modern Thought. Fontana Press, p. 917.

41 Bothamley J (2022) Dictionary of Theories: One Stop to More than 5,000 Theories. Visible Ink Press, p. 637.

42 Turnbull C (1973) Economy, morality, and anthropology. In Turnbull C (Editor) Africa and Change. Alfred A Knopf, pp. 15–43.

43 Van Beek WEA (1990) Harmony versus autonomy: Models of agricultural fertility among the Dogon and Kapsiki. In Jacobson-Widding, Van Beek WEA (editors) The Creative Communion: African Folk Models of Fertility and the Regeneration of Life. Uppsala Studies in Cultural Anthropology (15) - Acta Universitatis Upsaliensis, pp. 284–305.

44 Minkus HK (1984) Causal theory in Akwapim Akan Philosophy. In Wright RA (Editor) African Philosophy: An Introduction. University Press of America, pp. 113–147.

45 Balandier G (2020) Le Royaume de Kongo: Du XVIe au XVIIIe Siècle. Pluriel, p. 284.

46 Langley P (1976) Approche ethnolinguistique de l'environnement rural et son utilité pour l'aménagement. Environnement Africain, problèmes et perspectives. Institut International Africain, Dossier Spécial 1: 95–108.

47 Diamond JM (1999) Guns, Germs, and Steel: The Fates of Human Societies. W. W. Norton & Company, p. 480.

48 Diamond JM (2000) De l'inégalité parmi les sociétés: Essai sur l'homme et l'environnement dans l'histoire (Traduit de l'Anglais par Dauzat PE). Nouveaux Horizons, p. 695.

49 Wrangham R (2019) The Goodness Paradox: The Strange Relationship between Virtue and Violence in Human Evolution. Vintage Books, p. 400.

50 Vansina J (1992) Habitat, Economy and Society in the Central African Rain Forest. Berg occasional papers in Anthropology 1. Berg Providence/Oxford, p. 16.

51 Adams JS, McShane TO (1992) The Myth of Wild Africa: Conservation without Illusion. University of California Press, p. 282.

52 Dal Negro S (2004) Language contact and dying languages. Revue française de linguistique appliquée IX (2) : 47–58.

53 Dressler WU, Wodak-Leodolter R (1997) Language preservation and language death in Brittany. International Journal of Sociology of Language 12: 33–44.

54 Tempels P (1969) Bantu Philosophy. Présence Africaine, p. 190.

55 Mudimbé VY (1997) Tales of Faith: Religion as Political Performance in Central Africa. The Athlone Press, p. 231.

56 Jones DV (2010) The Racial Discourses of Life Philosophy: Negritude, Vitalism and Modernity. Columbia University Press, p. 231.

57 Tellier D (2011) Apprendre à philosopher avec Bergson. Ellipses Edition, p. 252.

58 Masolo DA (1994) African Philosophy in Search of Identity. Indiana University Press and Edinburgh University Press, p. 301.

7 Capacities, Institutional Arrangements, Democracy and Climate Change in Central Africa

Introduction

At the end of Chapter 6, two objections were thought to be leveraged against the argument that the views and voices of communities of Central Africa were not sufficiently taken into account in the process to make policies to conserve biodiversity and deal with climate change in Central Africa. The first rebuttal was that representatives from Central Africa were always present in discussions of globally important questions of the biodiversity erosion and climate change. The second objective was that the 'free, informed and prior consent' was always obtained from communities before any action was implemented. These arguments are sensible and needed to be discussed and put in global Central African perspectives in a constructed narration. This chapter intends and hopes to achieve the construction of this narrative. This chapter provides regional perspectives under which the critique about these objections is analyzed; the chapter also appraises the general conditions under which the policies to stop biodiversity collapse, to reduce the effects of climate change and mitigate the process of emitting greenhouse gases (GHGs) have been operationalized in Central Africa.

Institutional Capacities and Arrangements and Climate Change in Central Africa

Chapter 4 discussed the limitations of scientific capacities in Central Africa. Limited scientific capacity impacts the content quality nations of Central Africa bring to biodiversity conservation and climate change negotiations. Chapter 2 discussed low-quality scientific contents of what Central Africa produces. What has to be added is that often qualified academics (personnel) are often rejected to participate in biodiversity conservation and climate change events because of their country's politics; which is often heightened by the political fight between academics themselves. Internal academic power struggles[1] heightening the impact of political politics

DOI: 10.4324/9781003493754-8

is not unique to Central Africa, but saying that this lowers the scientific work's quality does no injustice to Central Africa. Limited institutional capacities and the ways in which the political institutions are arranged in Central Africa cause these drawbacks.. Not repeating what Chapters 2 and 4 already discussed, this chapter is about institutional capacities and political structure arrangements and their impact on the questions of biodiversity loss and climate change.

Present Central Africa political institutions were inherited from the colonial era. They were mostly inherited from Belgium and France. The Cabinda (Angolan Enclave) and Equatorial Guinea were Portuguese-colonized; north-western Cameroon was British-ruled and was later reunited with the Francophone Cameroon. Current Central Africa political institutional makeups are reminiscences of each country's colonial history.[2] This fact has massive implications. Similar colonial methods employed in the past continue to prevail in Central Africa. Broadly, the colonialism legacy is that the colonizers left, but the local tyrants who now colonize their own people[3] replaced them. Nuances necessarily play here; each country having followed its own historical path is expected to exhibit traits that deviate from this general new normalcy in some respects. Variability in how to mimic the colonial authority is implemented in different countries is often used complacently to argue that not all Central Africa's political leaders are equally as bad as some political analysts suggest. Surely, there are nuances but, generically, the political institutions in Central Africa have someone in the center of the political power[4] from which point every spoke leads to the outer society's layers. The power's geographic center moved from the capital of the former colonizing country to the capital city of the newly independent country. However, this power's geographic location moved but just half-way from the metropolis to the new capital city in the tropical equatorial forest. Also, political analysts maintain that the decolonized countries possess largely token economic power.[5] I claim that even the political power decolonized countries possess is also massively a token-power, which is overwhelming demonstrated by its dependence on the former colonizing power to continue to exist.

Throughout Central Africa, the political leadership doesn't only ape the colonial ruling model. Central Africa's political leadership doesn't only copy and paste the negative side of what they think was the way colonial authorities managed states they created. Central Africa's political leadership also takes the negative side of what it claims to be the political attributes chiefs had in the traditional societies. The *idea of chief*[6] is fashioned around the beliefs that the chief owns everything that he (she) distributes to whom he (she) wishes and as he (she) wishes. The chief, the traditionalist argument goes, is given the best part of the meat (if not the whole of it). Central Africa's modern political leadership evokes the same traditional authority's negative idea to justify its ruthless rule. The chief incarnates of the totality of political powers (executive, judiciary and the legislative) and the moral authority. The chief incarnates the divine will, if he isn't simply the divine. This forms an explosive power concoction at one single individual's hands, but chief's followers take his wills and wishes to be a carefully and well-thought political program to deliver the well-being of the communities across the country.

Because political leaders represent the divine will, effective executive, judiciary and legislative powers, economic structures, law enforcement bodies and cultural structures serve one political leader who sits in the political center. The political strong man distributes the economic largesse and life's joys to their cronies; the shares that go to non-followers are punishments and long-day forced seclusion behind the walls. Lucky opponents go to forced seclusion behind the walls; unlucky opponents are physically suppressed. In such an environment, state structures are either inexistent or very weak to implement any sensible sustainable development program with comprehensible objectives. This is why, for example, a climate change country[7] fragility study found that Central African Republic and the Democratic Republic of Congo were the most fragile countries. This study suggested that countries with solid political and social structures are the most resilient and the ablest to handle the direst conditions in a world where meteorological conditions will continuously change. The new weather patterns impose new situations, which need more local capacities to deal with. Countries where political and social structures are weak will be the weakest to deal with changing climate conditions. This is the case in countries whose political and social structures deviate from their roles of providing essential social services and infrastructures and redirect meager resources they have to keep up with the demands of strong men presiding over collective destinies.

Coping with climate change in Central Africa doesn't only demands more and new social and political structures. There is a need for the existing structures to function normally. Institutional capacities do still need to be built so that they can reach the level from which they can respond adequately to primary community needs they are supposed to serve such as education, and public assistance when needs arise, and healthcare.[8] They can be re-arranged to factor climate change in the processes aimed at serving communities. Central Africa's political structures need to de-focalize from essentially serving one single individual who politically hi-jacked their countries. If the climate change country fragility study[7] is right, vulnerability to climate change is more an issue of how political structures and social infrastructures are laid out and are readied to contribute to community resilience; it isn't just a question of how hardened climate conditions become. In countries in the Amazonian Basin, climate exacerbated existing infrastructure deficiencies and socioeconomic conditions [...] in impoverished areas[9] more than it did in urban areas where the wealthiest people live.

Climate Change in Central Africa and Democracy

A probably most important consequence of the queer mixture between the uneducated traditional idea of the chief and the inexpert imitation of what was apparent in the colonial power exercise is the suppression of voices[10] of the multitude that are outside the strong men crony circles. Suppressing the majority's voices constitutes a serious impediment in the way the political participation is organized in Central Africa. The powerless are rarely able to speak; when they shout out, their voices are rarely counted and they are least heard. When consulted, they are given the directions and leading questions to which they should respond. The politically powerful control the responses and lead to a semblance of unanimity. Powerless people are groups of different immensely marginalized groups; their clutches are moneyless and are less or lightly educated people. They include indigenous people, the large majority of women and the youth. Some qualified individuals, when they don't sing hymns to glorify the powerful big men, are also often purposely excluded and made extremely powerless and voiceless. They are pushed to socially secluded spots by the semi-literate people who grab political powers by undemocratic means. Public opinions count little in ways laws are made and enforced. Ideas are seldom discussed, and only the powerful groups' ideas count. Lacking true public participation in processes to make laws, to amend the ones to be amended and to enforce them, the mightiest ones forcefully impose unjust laws on communities. Law-making processes and law enforcement are abnormally flawed procedures. Carefully looking at these law-making processes and the routes to access to power in Central Africa, it is just to think that everything political is a masquerade democracy, if democracy can be said to have been tried at all. The masquerade democracy is exercising power by following the methods tyrannies, despotisms and autocracies use. Huguex[10] suggests that democracy is totally absent in Central Africa where elections are tricked and constitutions are changed frequently to protect the interests of people who hold the cultural, economic and political powers.

The lack of democracy and free, informed and prior consent oppose each other. This is not only conceptually but also practically when consent was sought and said to be acquired. The consent international organizations obtain when they seek it is very often the consent of the chief, no matter what level this authority lies. This is so because the *idea of chief* doesn't solely apply to the big strong men in the central power. The imaginary of the *idea of chief* cascades down to every level of authority; smaller influence circles form around traditional chiefs, intermediary chiefs mandated by the center of the political powers, school masters, immigration officers, etc. Traditional chiefs and smaller intermediary administrative authorities behave in the same ways as the big men who run Central African countries but in much smaller geographic radiuses. When the consent is sought from communities, chiefs speak on behalf of their communities without consultations. As traditional authorities or constitutional administrative authorities, chiefs behave similarly and exclude social consultations. Chiefs (modern or traditional) are considered to be the enlightened people with clairvoyance that gives them the moral authority to act on behalf of *their* communities. In such circumstances, the consents often thought being obtained from the majority of communities express what chiefs agree on. Clearly, this is far from being a free consent; chiefs shield the opinions of the majority to get out as they are expressed. Chiefs argue that there is a need to present something coherent, but coherence isn't the moral ground to suppress opinions that diverge from what the political leaders think is good.

The consent proclaimed in different biodiversity and climate Conference of Parties has to be from an informed community. Being informed equals having facts and being able to analyze them and draw conclusions. Having information makes participation one democracy's foundational pillar. However, very few citizens of Central Africa understand academic and political biodiversity and climate change jargons (Chapter 2). To say the least, scientific and political biodiversity and climate change jargons are complex and significantly lengthy to translate in local languages if they are to make sense. The novelty of the climate situation itself (Chapter 1) partly explains this difficulty; the modest education level in most of Central Africa is another explanation. The lower social strata people's majority isn't appropriately scientifically literate. Literally half-literate people abound in local Central African communities; half-literate people have little adequate climate knowledge. Stating isn't to be mean; it isn't to ignore the value of local traditional knowledge. This doesn't even say that the probability of meeting highly skilled people in these communities is null. The point is that local traditional knowledge is insufficient to grasp the entirety of the current climate and biodiversity crises (Chapter 2). The probability of encountering highly skilled scientists in the local communities is insignificant, even if it is still a possibility. Furthermore, Central Africa bears the characteristic of the inexistence of past true exercises of democracy caused chiefly by the history of suppression of freedom of speech. The colonial power and the national authorities that replaced the colonizers suppressed the freedom of speech. The long-repressed speech inhibited the capacity for most people to overtly give their voices when needed and when conditions to do so are genuinely democratic in essence. People in Central Africa would, for most time, hold back on their own opinions in the fears that one day they might be reprimanded by the power holders. This makes it very difficult to know even how people would express themselves, if they were given a true opportunity. The people of Central Africa are mostly not used to expressing their opinions openly visà-vis anybody; capacities to conduct long and constructed arguments remain mostly dormant in people.

With limited knowledge on concepts, idioms and the rationale behind climate change subjects, most communities in Central Africa accept some deals they are presented with sometimes because their children think to understand what the discussion was about; they convince their community to accept the action presented. Sometimes the deal presented is accepted just because one non-governmental organization takes its time to pounder one single point over and over again. Pestering people to repeat one point over and over again is a technique to ensure that the communities endorse something even if isn't essential for communities. Pestering people isn't open discussion; discussion is based on shared knowledge, a shared language's sophistication that eases the conversation with communities. Pestering people is all but a democratic learning exercise. The 'park as wildlife maternity' approach (Chapter 6) exemplifies the situations where complex issues are condensed to one facet while all other façades are hidden. In these circumstances, talking of free, informed and prior consent should be massively cautioned. Even if communities authentically sign the consent, it might just reflect the opinions of politically strongest people. The consent is also often obtained from partial comprehension of the information shared, at most. If serious community consent means for it to be free and informed, the road to get the authentic and carefully formulated consents from Central Africa is still very long.

Voices of local communities should be attended to in negotiations wherein the future of the planet is being decided upon (Chapter 6), but they aren't sufficiently taken onboard. Local sustainable development real needs are absent in schemes to create conditions for the sustainable use of natural assets. With this, conditions to fight climate change and preserve biodiversity are not being created. To address sustainable development, fight climate change and preserve biodiversity at the same time remain a difficult question to solve and will take a long time. One alleyway to reach optimal solutions is to have functional political democracies. Central Africa is less fertile for democracy because it lacks thought and expression freedoms. For climate change to become part of the Central Africa's political culture, political changes toward true participation have to happen in the first place. A public sphere[11] whereby democratic debate can happen unrestrictedly is to be constructed.

There is a counter-argument to the statement that political participation is still to be constructed in Central Africa. This rebuttal goes that the emergence of non-governmental organizations (NGOs) achieves greater participation without dramatically changing the political processes. Numerous NGOs' raison d'être is to increase the political participation and build the capacities of local communities to speak on behalf of constituencies and the economically and politically voiceless. NGOs are an undeniable fact in the social landscape of Central Africa. Irrefutably, NGOs make significant leaps toward inclusion and political participation. There are NGOs that do a laudably great job in the promotion of not only political participation but several critical issues. Chief among these critical issues has been climate change and biodiversity conservation. It would be foolish to ignore the positive contribution of these civil society organizations and, particularly, of the NGOs at all levels, in the advancement of these causes across Central Africa. Most biodiversity conservation and climate change work in Central Africa has begun with the work of several international NGOs, and that work continues to be done by structures of the civil society.

But NGOs Are Not Necessarily the Best in Climate Change

NGOs deserve such a eulogium, but it is intellectually inept to ignore the limits and dark side of some of these organizations. First, there are many more NGOs in Central Africa now than there were just a few decades ago. Most of them have the same objectives, follow the same methods and implement identical activities often in same communities. They all have the same sources of financial incomes that allow them to work. Second, drawing financial resources from the same sources leads NGOs to harshly compete over same limited funding. Competition over limited funding creates conflicts and makes it hard to have economy of scales that would allow NGOs to achieve a greater combined impact. Third, competition to gain the funds pushed large NGOs to create bureaucratic machineries to write funding proposals that neatly match donors' expectations. Proposal writing is a very expensive profession on its own right. Even though most often proposals are made rosy to please donors without taking into account field realities, they are nonetheless funded and proposal writing experts are highly paid on the money they generated. Fourth, reporting is now transmuted into a profession of its own; it recruits people who can play with words to emotionally move the donors. Moving the donors emotionally is done so sometimes (and only sometimes) in the absence of crude quantitative data. All reports NGOs write are not content-void; there are analytically good reports some NGOs produce. However, other NGOs are distressed about quantitative data to which they regularly oppose qualitative narratives, when they lack hard field-based data. NGOs' proponents of qualitative narratives ignore that qualitative analyses constitute important and sound scientific methods. Non-parametric statistics help put qualitative social data into different perspectives and sensitivities. Appealing to qualitative analyses doesn't mean to deviate from the epistemological demands of rigor, factuality and the repeatability of studies. Recourse to qualitative analyses isn't to narrate one's ideas, and it isn't an exercise of fitting records of rebellious field facts to what donors would want to hear.

Biodiversity and climate insufficient funds in Central Africa (Chapter 2) are often divided between NGOs that often hardly cooperate, except when they lobby donors to get more resources. Willing donors read what they want to hear, and NGOs produce glowing reports even when, sometimes, contents deviate from the field realities. In the end, when everyone expects NGOs to bring changes, there is very little change coming out of the NGOs' field actions.

Fifth, numerous international NGOs are globalized circuits to intercept financial flows that rich countries and the increasing world philanthropic industry channel to poor countries. NGOs capture significantly large international aid's portions by positioning themselves to be structurally more efficient than governments in delivering results. The scramble for funding is what drives international, national and local NGOs, but it isn't ignominious beyond repair. However, what boggles minds is that in most instances and at all scales, NGOs are equally inefficient as governments in Central Africa, and NGOS are also inefficient for similar reasons. Some international NGOs are so large (as are the governments in Central Africa); the greatest budget's parts are spent on payrolls and international, national and local operations. The paramount chunk of the scarce biodiversity and climate funds end up paying salaries and long-lasting field trips to hotels in towns (e.g., Goma, Bangui, Douala and Franceville). Ironically, these towns aren't the field for the citizens of Central Africa. The idea here is not to suggest that NGOs shouldn't pay the experts they recruit at the right level but to show that NGOs can skim themselves slim to get the work done where the true field is located.

National and local NGOs are often filled with incompetent personnel mostly recruited among friends and from family members (cousins, nephews and in-laws[12]). National and local NGOs share this feature with Central Africa's governments. Some international agencies also share this feature but in magnitudes incomparable to what happens with local and national NGOs and governments. Political elites create NGOs to achieve their political ambitions.[13] In some cases, NGOs are transformed into political parties; civil society leaders often use NGOs as political launching pads. Hence, inadequate climate funds fall prey to inexpert hands and are divided among family members and friends and are used to politically promote the elites to achieve political goals in Central Africa.

Capacity Building for Central African Citizens Has to Radically Become Different to Work

A final example that merits citing here is the capacity building notion imbedded in climate change projects in Central Africa. Nearly all international NGOs talk about and do some work to build capacities[14] for Central African nations to enable them to navigate through future demands. International global sustainable development NGOs, humanitarian NGOs, biodiversity conservation NGOs and climate change NGOs all practice some capacity building activities. International NGOs support most people currently doing research in Central Africa, one way or another. I benefited from that unquestionable support to study abroad; I know how some international NGOs try hard to build capacities in Central Africa.

Notwithstanding these recent-year great strides in some Central African countries, the narrative of capacity building has limited impact on ground to the point that sometimes it itches to sit down where Central African human capacities are discussed. A comprehensive scientist listing exercise

(Chapter 3) indicated that 62% of researchers who publish on the Congo Basin reside outside of the Basin itself. Clearly, this reality expresses a great need to transfer skills housed outside of Central Africa to local residents for communities to own knowledge scientific research produces on the Congo Basin. Academic capacity building for local citizens is still critical despite years of investment. Less-sanguinely phrased and against expectations, years of capacities building have not borne abundantly. Stating this has nothing to do with a romantic pan-African nationalistic revolutionary spirit; it simply means that there are lessons to be learnt from the three-decade capacity building experience. Institutions building capacities in Central Africa should learn from past failures and successful models that worked in other world regions.

There are reasons for the meager results the investment made to build national capacities yields in Central African countries. To begin with an obviously internal Central African paradigm, most highly trained Central African citizens choose to immigrate to countries where they can get better paid jobs after they complete degrees in Western universities. Surely, better paid jobs they get in Western countries come with clear career paths that ensure highly trained Central African citizens to socially move up. The other detail of lacking expertise in the Congo Basin is that employing NGOs send out limited Central African people to achieve elevated degree training; some employing NGOs limit the capacity building offers to short field training. Short field training courses allow employees to be just good data collectors to support highly skilled international employees to publish. Some Central African citizens who are luckily supported by their employing NGOs fall out before completing their curricula; dropping out before completing degrees is natural, not everyone who is enrolled at a university finishes the training.

The largest portion of funding to support human capacity building in Central African countries is also given to young students from countries donating funds for climate change activities. Donor country's young expatriates come to the Congo Basin for doctoral research; they leave once they complete their research, leaving little knowledge behind. Without naïve, narrow and romantic African nationalistic revolutionary ideologies, it is no wonder expatriates possess most important knowledge on Central Africa, lodged in Western institutions. Inadequate climate change funds allocated to Central Africa to build local capacities finally build donor countries' capacities to solve Central African problems. René Monory, French Senate's president (1992–1998), said in a political emission Canal France International broadcasted (19 February 1995), that one solution to decrease the France's unemployment was to export French expertise.[15] This statement says that capacity building generally reflects political power. Training experts, beyond the genuine will to create knowledge, is a political act.

Does suggesting that capacity building generally reflects political power imbalances mean that current Central Africa's capacity building schemes follow well-thought plans to keep Central Africans playing a second-order role while the true expertise remains in donor countries that should continue to lead? This inference is difficult to swallow wholly; it's even difficult to assert what René Monory thought entirely. However, some Central African voices see in the political power imbalances between Western powerful nations and countries in Central Africa some willingness by powerful nations to maintain upper hands on the training agendas. There is a large Western-skewed imbalance of numbers of people benefiting from capacity building in Central Africa. There are more Western citizens trained on and in Central Africa; comparably, only meager numbers of people from local communities are trained. Imbalanced capacity building epitomizes, in other ways, the uneven distribution of economic and political power in the world.

Are There Reasons to Hope for Central Africa and the World?

With all the narratives above (Chapter 3), is there any hope for Central Africa to survive through the changes happening currently? This question is worthy being asked when internal, national, institutional and individual capacities of Central Africa are weak. Human capacities provide keys to sustain the efforts to fight against and adapt to the effects of climate change. The question deserves to be asked because international, national, mid-level and local non-governmental organizations are inadequately, both conceptually and practically, equipped to help people of Central Africa fight against and adapt to climate change. Certainly, it would be morbid thoughts saying there is no hope at all. Several factors present in Central Africa itself impose weighted optimism. The fundamental reason for this moderate optimism is that people realize that they shouldn't expect abrupt, profound, global and immediate metaphysical transformation[16] that would change their life for better (Chapter 1). People understand that neither governmental structures nor international aid will sort the totality of the climate change problems. If any miracle, there will be only one: the return back to the natural order of things.[17] The natural order of things implies communities to work out their own ways to adapt to new weather patterns. People in Central Africa are always resilient against adversity. No matter how dire adapting to climate change is, communities in Central Africa will adapt. Large adaption plasticity helps communities in Central Africa survive through hard times; adaption plasticity is a regional culture. Being resiliently self-reliant doesn't mean being aid-repellent and cooperation-repulsive, but the supportive foreign hand will not do all the climate adaptation work and helicopter-deliver results for Central Africa.

To use the adaptation plasticity to adapt, political management structures must change in how they manage collectivity culturally, economically and politically. Change required should be more than cosmetic; superficial changes are often more complicated and lack impacts. Impactful changes in management should be deep and mind-shaking; in some areas, they are to be revolutionary. Deep and mind-shaking changes need long time before fruiting. However, strong and long-lasting support to improve governance is what Central Africa expects from the international community to help them embrace climate change with sufficient chances to succeed. Strong support to improve public policy governance is the only way to help people cope with climate change. Better governance ensures that global climate change measures are implemented irrespective of who runs the show. But better governance also needs *competent* leadership. The support Central Africa needs is the help for a new political culture to emerge. A new political culture is needed! It is needed more than any technical climate solution.

The second reason for a restrained hope is that *Homo sapiens* are capable of changing habits. Even the most brain-ingrained habits can be dropped in front of life-threatening dangerous circumstances (Chapter 1). How people adopted all measures to avoid spreading covid-19 across the world is an example of such an abrupt adaptation to life-saving requirements. Taking such measures require people to feel directly threatened in their own lives. So far, climate change doesn't seem to feel like a direct and individual life threat. Significantly, however, common Central African citizens feel climate change effects in their lives. Heavy floods are now normal, but they occurred in several decadal sequences in the past. In some Central African countries, prolonged droughts commonly occur now where they were previously infrequent. Floods and droughts take economically high tolls from poor people; they are both costly in human lives and conduce to people movements principally within Africa[18] and, marginally, to the outer world. These climate change effects affect communities; they are likely, even if moderately, to raise the opinions and push the governments act drastically. One can only hope that this will not wait more millions people to die or become either internally or externally displaced!

The final reason to be modestly hopeful is the global world civil society's emergence. The global civil society offers a life ideal to the Central Africa's youth; it supports social movements in countries where demonstrations are often crushed. The global civil society is marred with problems specific to its own nature. But this is not specific to the global civil society because all human endeavors have their own problems to sort out before achieving significant impacts. Ideologies that drove most people across the world in the late 1960s vanished and left a vacuum where they once presented powerless people (youths, women and moneyless) with hope horizons and with reasons to politically fight.[19] However, the globalization is leading to the emergence of a worldwide consciousness through global social movements. Social movements amalgamate similar demands for better governance[20] and care for nature. Assuredly, social movements are not always politically neutral. Most often, they are even politically motivated. Some social movements are even anarchical and destructive. Some social movements may even seem to be or truly become major hurdles to achieve sustainability, justice and social cohesion.[21] However, despite these flaws, social movements hold a huge potential to pressure governments[22] world-widely to act on climate change, provided that they follow the same objectives and act constructively. Climate social movements are still rudimentary in Central Africa. The youth's apathy vis-à-vis climate change in Central Africa is fully understandable, considering how depressive poverty levels are in this region. High poverty levels make most young people devote the best of the energy they possess in the struggle to find food, clean drinking water and clothes, which constitute basic human needs. The lethargic situation in which the youth of central Africa dwell in will not continue for so long; young people in Central Africa are connected with the world via information technologies and build hopes and desires to live under conditions that youths in other parts of the world live in. This will increase the youth's political awareness across Central Africa. Hopefully, youths will press their political leaderships more than it is currently the case to invest national resources and efforts toward a better world where the issues of climate change and other primary necessities will be addressed with the care they deserve.

Notes

1 Lumby J (2019) Leadership and power in higher education. Studies in Higher Education 44 (9): 1619–1629.
2 Griffiths ILL (1995) The African Inheritance. Routledge, p. 216.
3 Bayart JF (2006) L'Etat en Afrique : Politique du ventre (Nouvelle Edition). Fayard, p. 439.
4 Jackson RH, Rosberg CG (2023) Personal Rule in Black Africa: Prince, Autocrat, Prophet, Tyrant. University of California Press, p. 328.
5 De Rivero O (2003) Le mythe du développement (Traduit de l'Espagnol par Robitaille R). Enjeux Planète, p. 241.
6 Cohen JH (2016) L'esprit de l'homme fort Africain: Mémoires d'un diplomate (Traduit de l'Anglais par Seiler C). Mediaspaul, p. 222.
7 Marcantonio R, Javeline D, Field S, Fuentes A (2021) Global distribution and co-incidence of pollution, climate impacts, and health risk in the Anthropocene. PLoS ONE 16(7): e0254060.
8 Kula N, Haines A, Fryatt R (2013) Reducing vulnerability to climate change in Sub-Saharan Africa: The need for better evidence. PLoS Med 10 (1): e1001374.
9 Andrade LdMB, Guedes GR, Noronha KVMdS, Santos e Silva CM, Andrade JP, Martins ASFS (2021) Health-related vulnerability to climate extremes in homo-climatic zones of Amazonia and Northeast region of Brazil. PLoS ONE 16 (11): e0259780.
10 Huguex V (2012) Afrique: Le mirage démocratique. CNRS Editions, p. 63.
11 Habermas J (2019) The structural transformation of the public sphere: An inquiry into a category of bourgeois society. Polity Press, p. 301.

12 Souop K (2015) Mort de la tribu originelle et refondation de l'Etat africain. In Djateng F, Kayser C (Editeurs) Les sociétés civiles en Afrique. Cahiers du Mapinduzi 4 : 60–69.

13 Mana K (2015) Société civile en République Démocratique du Congo Entre pathologies mortelles et construction d'un pouvoir intelligent. In Djateng F, Kayser C (Editeurs) Les sociétés civiles en Afrique. Cahiers du Mapinduzi 4 : 50–59.

14 Bracken U (2015) Contribuer au changement social? Le potentiel et les limites des projets de développement. In Djateng F, Kayser C (Editeurs) Les sociétés civiles en Afrique. Cahiers du Mapinduzi 4 : 76–91.

15 France 2 (1995) Heure de vérité – Emission télévisée de politique Française, créée et animée par François-Henri de Virieu. The emission was broadcast to Central Africa by Canal France Internationale. The emission cited here was that of 19 February 1995. The other two solutions René Monory proposed were (1) to make enterprises become a place where training would be continuous while reducing the social costs and (2) for France to become economically more present in the new oriental economic powers (countries).

16 Welburn D (2018) Rawls and the Environmental Crisis. Routledge, p. 146.

17 André Fossard was interviewed on 05 November 1993 at the literary emission called 'Jamais sans mon Livre', which was broadcast to Central Africa by Canal France Internationale.

18 Beauchemin C, Ichou M (2016) Au-delà de la crise des migrants: Décentrer le regard. Karthala, p. 198.

19 Hollande F (2019) Les Leçons du pouvoir. Editions Stock, p. 500.

20 Smith B (2008) Social Movements for Global Democracy. The Johns Hopkins University Press, p. 286.

21 Morell IA (1999) Emancipation's Dead-end Roads? Studies in the Formation and Development of the Hungarian Model for Agriculture and Gender (1956–1989). Acta Universitatis Upsaliensis, Studia Sociologica Upsaliensia 46, p. 485.

22 Graubart J (2008) Legalizing Transnational Activism: The Struggle to Gain Social Change from NAFTA's Citizen Petitions. The Pennsylvania State University Press, p. 170.

Index

Note: Page numbers in *italics* and **bold** refer to figure and tables, respectively.

Adorno, Theodor 70
African traditional knowledge 26–27
Annan, Kofi 28
Anthropocentrism 55–56
anthropocentrism 55
Anthropogenic GHG emissions 6
Aquinas, Thomas 79–80
Arrhenius, Svante 5

biocentrism 55
biodiversity conservation 75, 78–79,
    91–92; Central Africa's voice
    83–89; institutional capacities and
    arrangements for 96–98; park, as
    maternity space for wildlife species
    89–91; reading back-to-back 82–83;
    Western-conceived conservation
    biology 82
biodiversity erosion 10, 75;
    implications 76
biodiversity reflection framework 77
bio-engineering 20
biologism 81
biophilia *qua biophilium* 79
biophilism 78
biophysical environment 88

carbonated economy 9
Central Africa: biodiversity
    conservation 83–89; capacity
    building notion 103–105;
    citizens' love of nature 86;
    climate change challenges 3–4, 8,
    99–102; communities' per capita
    CO_2 emissions 70; concept of
    sustainability 10; coping with

climate change 26–27; future
    105–107; interpretation of
    precautionary principle 35–36;
    preservation of biomes 28; right of
    the citizens 70; social movements
    106–107; theological traditions 86
Central Africa's political leadership
    and political institutions 96–98;
    corruption 97; idea of chief 97–99;
    non-governmental organizations
    (NGOs) 101–104; role in coping
    with climate change 98–102
Chevrolets 69
climate change negotiations 8, 36–37
climate diplomacy 36–37
climate-induced calamities 4;
    deviations in weather patterns
    16–17; emergency chain reactions 7;
    Holocene and Anthropocene periods
    4; human collective memory and
    17–18; industrialization and 5–6;
    technological-based solutions to
    19–21; traditional knowledge for
    coping 26–27
climate payments 33–35
climate resilience 41
climatic skepticism 25–27
Climatology 2
CO_2 emissions 17, 63–65, *64*, **65**; CO_2
    leaking notion 69–70; CO_2 seller
    countries 68; per capita emissions
    64–65; strategies to mitigate
    66–71; super emitter countries 64;
    technicalities of CO_2 quantification
    68; trading provisions 67
collective climate consciousness 13

Conference of Parties (COP) 1, 14–15, 36, 39; obstacles to achieving progress on climate change 25; operational objective of 24
Congo Basin 9; climate solutions framework 57–60; major CO_2 emitters 65; research and innovation investments 69; roles and moral responsibilities in climate action 40–41
consumerism 52
consumption ideology 53
'copy and paste' culture 58–59
Covid-19 crisis: emergency measures for dealing 18–19; negative side and positive side related to 21; vaccine development 19
Covid-19 pandemic 7
cultures 15, 56; scientific culture 81

Darwinism 52
decarbonizing economies 7, 9, 73
Democratic Republic of Congo (DRC) 1, 8, 33, 42, 46–48; auction of oil and gas blocks 29, 35; carbon-dependent economic growth 60; climate adapting and mitigating strategies 60; contextualized climate knowledge 47; population growth 57, 59; roles and moral responsibilities in climate action 40–41
Diamond, Jared 87
dissipation time 14
Dworking, Ronald 81

ecologism 92
Élan Vital 80
environmental determinism 87

financial assets in sustainable development 47
financial mechanisms to fight climate change *31*, 31–33; sustainable development plans 46–48; transfer of climate change fund 33–35; USD fund 33
food production 53
Fords 69
fossil fuels trade 7
Francis, Pope 83
Frogneux, Nathalie 11
functional and functioning democracy 10

geo-chemical-physical climate change causes 51
geo-engineering 20
Gilroy P 84
glaciations 6; forests retracted during 6; in Holocene period 5–6
global climate change 13
greenhouse gases (GHGs) 1, 17, 24, 32, 50, 63, 67–68, 96; consequences of 7
greening economic activities 67–73

Harris, Annabelle 11

industrial revolution 13
instantaneity trap 54
intergenerational justice 54
Intergovernmental Panel on Climate Change (IPCC) 6
international climate finance *see* financial mechanisms to fight climate change

Kahuzi-Biega National Park 88

land use planning process 9
Léopold, Aldo 57, 82
Loss and Damage Mechanism (LDM) 7, 32–33
Lovelock J 78

Monory, René 104–105
moral responsibilities in climate actions 40–41, 69
Morris RC 84

national technical body, significance of 42; strategic and specific objectives 42–46
naturalism 55
Natura naturans 56
Natura naturata 56
nature-nurture debate 56
no co-responsibility rebuttal 31, 34
non-governmental organizations (NGOs) 101–104

park, as maternity space for wildlife species 89–91
population growth 57
populist ideologies 26
precautionary principle 35–36

qua biophilium 86

re-enchanting the village 58
Ricoeur, Paul 54
Rio de Janeiro Agreement, 1992 1, 13, 27, 35
Rwandan economic model 59

Sancta simplicitas 70–71
Sartrean existentialist philosophy 51, 51–52
Scandinavian countries 58
School of Management of the Catholic University of Congo (SM-UCC) 2
Singer, Peter 41
social inertia 14; public indoctrination and 15
social safety net 58
Soulé, ME 83
Spinoza, Baruch 80
Spivak GC 84
sustainability matrices 41

technological-based solutions to climate change 19–21

UN 1992 Conference on Sustainable Development 3
UN Declaration on the Right to Development (UNDRD) 28
United Nations (UN) Framework Convention on Climate Change 1
United Nations Framework Convention on Climate Change (UNFCCC) 24, 40; precautionary principle 35–36; price action against climate change *31*, 31–33; principle of common but differentiated responsibilities 30–31; right to development 28–30

vitalism 79–83

Western industrialization 9–10, 15
Wilson, Edward O. 78
Wilson and Lovelock 81
Wilson EO 78

Yellow Vests 5, 54

Printed in the United States
by Baker & Taylor Publisher Services